T0224563

Evolved Morality:
The Biology and Philosophy of Human Conscience

Evolved Morality:
The Biology and Philosophy of Human Conscience

Edited by

Frans B.M. de Waal, Patricia Smith Churchland,
Telmo Pievani and Stefano Parmigiani

BRILL

Leiden · Boston
2014

Library of Congress Control Number: 2014931293

ISBN: 978 90 04 26387 1 (hardback), 978 90 04 26816 6 (paperback)

Printed in The Netherlands

CONTENTS

The page numbers in the above Table of Contents and in the Index refer to the bracketed page numbers in this volume. The other page numbers are the page numbers in Behaviour 151/2–3. When citing a chapter from this book, refer to Behaviour 151 (2014) and the page numbers without brackets.

[When citing this chapter, refer to Behaviour 151 (2014) 137–141]

Foreword

Evolved morality: The biology and philosophy of human conscience

Frans B.M. de Waal [a,*], **Patricia Smith Churchland** [b], **Telmo Pievani** [c] **and Stefano Parmigiani** [d]

[a] Living Links, Yerkes National Primate Research Center and Psychology Department, Emory University, Atlanta, GA, USA

[b] Department of Philosophy, University of California, San Diego, La Jolla, CA 92093, USA

[c] Department of Biology, University of Padua, Via U. Bassi 58/B, 35131 Padua, Italy

[d] Department of Evolutionary and Functional Biology, University of Parma, Parma, Italy

*Corresponding author's e-mail address: dewaal@emory.edu

The last decade has seen a renewed interest in evolutionary approaches to morality. In the 1970s and 1980s, morality and evolution were considered largely incompatible, even by biologists, but we are now returning to the view that morality requires and probably has an evolutionary explanation. This view is not without its controversies, however, hence the intense debate about moral origins within biology and philosophy. Since psychologists, anthropologists and neuroscientists have joined this debate, and also religious scholars are involved, it has become a truly interdisciplinary effort, which is reflected in the composition of the present volume.

Morality is often defined in opposition to the natural 'instincts', as a tool to keep those instincts in check. This is reflected in Dawkins' (1976) advice to teach our children altruism, since our species is not naturally inclined to such behaviour, and the even more extreme position of Williams (1988), who complained about nature's 'wretchedness'. Williams (1988, p. 180) felt that calling nature 'amoral' or 'morally indifferent', as Thomas Henry Huxley (1894) had wisely done, was not strong enough. He accused nature of 'gross immorality', thus becoming the first and hopefully last biologist to infuse the evolutionary process with moral agency.

By holding up human morality as an exception to the larger scheme of nature, these authors deviated dramatically from Charles Darwin's (1871) proposal in *The descent of man* that human moral behaviour is continuous with the social and affectionate behaviour of other species, and evolved to enhance the cooperativeness of society. Darwin saw morality as part of human nature rather than as its opposite, a view widely supported by new interdisciplinary evidence. Experiments on human infants indicate that moral understanding starts early in life, studies of our primate relatives (Figure 1) show signs of moral building blocks, psychologists suggest 'gut' judgments that arrive more rapidly than reasoning and logic, and anthropologists have documented a sense of fairness all over the world. We are moving towards a

Figure 1. An adolescent male bonobo appears lost in thought. Photograph by Frans de Waal.

Figure 2. Some of the speakers at the conference held in June 2012 at the Ettore Majorana Foundation and Centre for Scientific Culture (EMFCSC) in Erice, Italy. Top, from left to right: Jeffrey Schloss, Richard Joyce, Liane Young, Owen Flanagan, Patricia Churchland, Philip Kitcher, Simon Blackburn, Melanie Killen and Stephen Pope. Bottom: Stefano Parmigiani, Frans de Waal, Telmo Pievani, Darcia Narvaez and Pier Francesco Ferrari. Unknown photographer.

new appreciation of the roots of human morality as well as the brain structures that support it.

For this reason, we brought together a collection of experts from disparate fields with the goal of exploring how and why morality may have evolved, how it develops in the human child, how it is related to religious beliefs, and whether neuroscience and evolutionary theory can shed light on its functioning in our species. This being the topic of several recent books by some of our participants (Harman, 2009; Churchland, 2011; Kitcher, 2011; de Waal, 2013; Norenzayan, 2013), we felt confident that a lively debate would ensue. We did not shy away from the age-old facts-versus-values debate, also known as the 'naturalistic fallacy,' as attested by several contributions that explicitly tackle this complex issue. The implications of this debate are enormous, as they may transform cherished views about our status as a moral species, a species able to produce conscious moral judgments.

The workshop idea originated during Frans de Waal's visit to Parma just after another Erice workshop entitled *The primate mind* (de Waal & Ferrari, 2012). Stefano Parmigiani was quite confident of a positive response to a workshop proposal on the biological roots of morality and ethics. The workshop was eventually held in the beautiful Sicilian city of Erice, in Italy, from 17 through 22 June, 2012, under the auspices of the Ettore Majorana Foundation and Centre for Scientific Culture (EMFCSC) and the International School of Ethology. The EMFCSC was founded by the Italian physicist Antonino Zichichi, who acts as the Centre's President. Since its founding the Centre, now embracing 123 schools, has represented an important meeting place for scientists all over the world. Our meeting was part of the International School of Ethology directed by Danilo Mainardi and Stefano Parmigiani.

The participants (Figure 2) and many students and scientists in the audience strolled the pre-medieval streets of Erice, which city offers a gorgeous mountaintop view of the Strait of Sicily, and discussed the lectures afterwards in local restaurants. All in all, it was a most stimulating meeting, which we hope is reflected in an equally stimulating volume. All manuscripts were subjected to peer-review. For ease of reference, papers on related themes have been grouped together.

Acknowledgements

We are deeply indebted to Antonio Zichichi (President, Ettore Majorana Foundation and Centre for Scientific Culture-EMFCSC) for including the workshop as part of the 2012 program of activities. The editors are grateful to all hard-working referees, and thank the publisher for making this Special Issue of *Behaviour* possible.

References

Churchland, P.S. (2011). Braintrust: what neuroscience tells us about morality. — Princeton University Press, Princeton, NJ.
Darwin, C. (1981 [1871]). The descent of man, and selection in relation to sex. — Princeton University Press, Princeton, NJ.
Dawkins, R. (1976). The selfish gene. — Oxford University Press, Oxford.
de Waal, F.B.M. (2013). The bonobo and the atheist: in search of humanism among the primates. — Norton, New York.
de Waal, F.B.M. & Ferrari, P.F. (2012). The primate mind: built to connect with other minds. — Harvard University Press, Cambridge, MA.
Harman, O. (2009). The price of altruism. — Norton, New York, NY.
Huxley, T.H. (1989 [1894]). Evolution and ethics. — Princeton University Press, Princeton, NJ.

Kitcher, P. (2011). The ethical project. — Harvard University Press, Cambridge, MA.

Norenzayan, A. (2013). Big gods: how religion transformed cooperation and conflict. — Princeton University Press, Princeton, NJ.

Williams, G.C. (1988). A sociobiological expansion of "Evolution and Ethics". — In: Evolution and ethics (Paradis, J. & Williams, G.C., eds). Princeton University Press, Princeton, NJ, p. 179-214.

Section 1: Evolution

[When citing this chapter, refer to Behaviour 151 (2014) 145–146]

Introduction

What better way to start this Special Issue of *Behaviour* on evolved morality than with the altruism debate that has been raging in biology for nearly two hundred years? Not that the morality question can be reduced to altruism, but it is hard to deny that the two are related. Owen Harman delves in the history of this debate, and the question whether an evolved morality is even possible. Charles Darwin definitely thought so, but his contemporary and supporter Thomas Henry Huxley had serious problems with the idea. From the start, therefore, there have been contrasting opinions.

Major figures in biology tackled the evolution of altruism, from Fisher to Price and from Kropotkin to Williams. It is a rich history that revolves around what biologists call the 'puzzle of altruism'. At first sight, it seems hard to account for the evolution of behavior that is costly or risky to the organism. Biologists have concluded, however, that not only is the genetic evolution of altruism possible, it is highly likely under certain circumstances. Proof is found in its widespread occurrence.

The same question of selection for altruism is taken up by Christopher Boehm, but now focused on the human case. As an anthropologist, Boehm has extensively studied the egalitarianism of human groups and the social pressures that shape behavior. Instead of looking at kin or group selection for an answer, he documents social selection, an underappreciated phenomenon. Human groups do not take kindly to deviant behavior, and have ways of expelling or eliminating individuals perpetrating it. In the long run, this may affect gene distribution. As a complement to reputation building, social selection may profoundly shape human conscience.

The final contribution to this section is by Frans de Waal, who explores the much-debated Is/Ought divide, often used as an argument against evolutionary accounts of morality. Instead of viewing animal behavior as a mere manifestation of how things are in nature, without any bearing on how things ought to be, de Waal argues that animal behavior follows goals and ideals, hence is fundamentally normative. Animal behavior may not be normative in the same sense as the distinction between right and wrong, but it is clearly infused with actively pursued values. As such, it is not as far removed from human normativity as often assumed.

The Editors

Review

A history of the altruism–morality debate in biology

Oren Harman *

Department of History and Philosophy of Science, Bar Ilan University,
Ramat-Gan 5290002, Israel
*Author's e-mail address: oren.harman@gmail.com

Accepted 22 August 2013; published online 27 November 2013

Abstract
Many different histories of the altruism–morality debate in biology are possible. Here, I offer one
such history, based on the juxtaposition of four pairs of historical figures who have played a central
role in the debate. Arranged in chronological order, the four dyads — Huxley and Kropotkin,
Fisher and Emerson, Wynne-Edwards and Williams, and Hamilton and Price — help us grasp
the core issues that have framed and defined the debate ever since Darwin: the natural origins of
morality, the individual versus collective approach, the levels of selection debate, and the Is–Ought
distinction. Looking forward, the continued relevance of the core issues is discussed.

Keywords
altruism, morality, evolution, history.

1. Introduction

The existence of altruism in nature has preoccupied biologists ever since
Darwin. Uncovering myriad examples across taxa in all kingdoms, biologists
have wondered how such phenomena could be possible if evolution is a game
of survival of the fittest. Like Darwin before them, they felt hard-pressed to
find solutions that would chime with the tenets of evolutionary theory.

For Darwin the mystery that needed explaining was how such diverse behaviors and morphologies, as were observed in ants, had arisen and were
maintained. After all, a given species — the leaf-cutter ants of South America, for example — had castes that differed in weight up to three hundredfold,
from miniature fungus gardeners to giant soldiers. If such workers left behind
no offspring, how could natural selection be fashioning their traits through

their own direct kin? The queen and her mate had to be passing on qualities through their progeny — massive heads, gardening scissor teeth, and
bewildering altruistic behavior — they themselves did not possess. It was a
problem related to heredity, and anything but trivial: How could traits, both
of form and behavior, perform such Houdini acts in their journey from generation to generation? Darwin called it "by far the most serious difficulty,
which my theory has encountered" (Darwin, 1859).

But it also happened to be a glaring exception to "nature, red in tooth
and claw" (Tennyson, 1850). Virgin nursemaids made little sense if success
in the battle of survival was measured by the production of offspring. After
all, as Darwin himself wrote, the very essence of evolution was that "every
complex structure and instinct" should be "useful to the possessor"; natural
selection could "never produce in a being anything injurious to itself, for
natural selection acts solely by and for the good of each" (Darwin, 1859). It
seemed illogical.

Ultimately, Darwin solved the conundrum to his satisfaction by positing
the notion of a 'community': Who benefits from the toil of the nursemaid,
the forager, and the soldier? The queen, who would be free to concentrate on
procreating — and by extension the entire growing family. This idea would
one day be called 'group selection' and excite enormous debate. But for the
meantime, Darwin soon came to believe that it played a crucial role in a
much grander drama: "The social instincts", he wrote in *The Descent of
Man*, braving a more sensitive subject, "which no doubt were acquired by
man, as by the lower animals, for the good of the community, will from
the first have given him some wish to aid his fellows, and some feeling
of sympathy" (Darwin, 1871). Evolution was the key to the beginnings of
morality in humans.

And so, from the outset, the problem of altruism and the problem of
morality were joined to each other in biology. To unfurl their entanglement,
many different histories could be written, in principle and practice alike,
each depicting a unique 150-year intellectual tale. In what follows, I will
attempt to sketch one such history by positing four individuals against each
other, arranged chronologically so as to paint a developing story. We will
meet T.H. Huxley and Peter Kropotkin, R.A. Fisher and Alfred Emerson,
Vero Cope Wynne-Edwards and George Williams, Bill Hamilton and George
Price. In some cases, the dyad represents two distinct sides of a theoretical
debate, in opposition to each other; in others, the juxtaposition is meant to
describe the development of a particular idea, from its early understanding

to a more mature treatment. Often the interlocutors were directly addressing each other; sometimes they did not know each other — only each other's ideas. But arranged in these sets, each dyad helps to throw light on one of four central themes of the historical altruism–morality debate. These are: the question of the natural origins of morality, the individual versus collective approach, the levels of selection debate, and the Is–Ought distinction. Taken together, these four themes illuminate the fate of Darwin's 'most serious difficulty', and help us better understand the current debates about altruism in biology.

2. Evolution of morality

Aboard the H.M.S. Rattlesnake, like Darwin on the Beagle before him, Thomas Henry Huxley traveled to the southern seas, encountering nature in the hullabaloo of the tropics (Desmond, 1994). It was a glorious clamor, indeed, but what philosophical lessons did it harbor for man? Years after his return, already famous and known to some as 'Darwin's bulldog', Huxley offered his thoughts in a piece titled 'The Struggle for Existence in Human Society: A Programme'. Imagine a wolf chasing a deer, he asked of his readers. Had a man intervened to aid the deer we would call him 'brave and compassionate', as opposed to 'base and cruel' were he to aid the wolf. But this was a confusion. Under the 'dry light of science', neither action was more admirable than the other, the ghost of the deer no more likely to reach a heaven of 'perennial existence in clover' than the ghost of the wolf 'a boneless kennel in hell'. "From the point of view of the moralist the animal world is on about the same level as a gladiator's show", Huxley wrote, "the strongest, the swiftest, and the cunningest... living to fight another day". There was no need for the spectator to turn his thumbs down, "as no quarter is given", but "he must shut his eyes if he would not see that more or less enduring suffering is the need of both vanquished and victor" (Huxley, 1888).

Christians imagined God's fingerprints on nature, but it was the meddling of the Babylonian goddess Ishtar that seemed to Huxley more to the point. A blend of Aphrodite and Ares, he wrote in his 1888 essay, Ishtar knew neither good nor evil, nor, like the Beneficent Deity, did she promise any rewards. She demanded only that which came to her: the sacrifice of the weak. This was a matter of necessity, not morals. For Darwin morality had come from the evolution of the social instincts, but to Huxley instincts were antisocial — their primeval lure the bane of man's precarious existence. However much man searched for it there, nature could provide for him no moral compass.

How fortunate, then, that nature had played a kind of trick. For in us she had fashioned quite literally a walking paradox, a "strange microcosm spinning counter-clockwise", Huxley called it. The human brain, after all, a product of evolution as much as the feathers of the peacock and the tentacles of *Physalia physalis*, could transcend the natural imperative, replacing indifference and necessity with caring and ethical progress. This would be "an artificial world within the cosmos", a civil and cooperative counterweight to the pull of bloody battle — "an audacious proposal", perhaps, but in our reach nonetheless. Self-restraint oiled by education and law would prove the negation of the struggle for existence, man's glorious rebellion against the tyranny of need. Altruism may not be a part of nature, but this was man's "nature within nature", sanctioned by his evolution. Morality was a uniquely human affair (Huxley, 1898).

To the north and east, in Russia, Price Peter Kropotkin demurred. As a youth he had traveled to Siberia, his own form of the voyage of the Beagle and Rattlesnake, and encountered a nature rather different from Darwin's and Huxley's (Todes, 1989; Dugatkin, 2011). Contrary to Huxley's tortured plea to wrest civilized man away from his savage beginnings, he believed, it was rather the return to animal origins that promised to save morality for mankind. In a direct reply to the English morphologist, Kropotkin published a series of five articles between 1890 and 1896 that in 1902 would become known as the book *Mutual Aid*. Here Kropotkin sank his talons into "nature, red in tooth and claw". For if the bees and ants and termites had "renounced the Hobbesian war" and were "the better for it" so had shoaling fish, burying beetles, herding deer, lizards, birds, and squirrels. "Wherever I saw animal life in abundance", he remembered his adventures in the Afar, "I saw Mutual Aid and Mutual Support" (Kropotkin, 1902).

This was a general principle, not a Siberian exception. There was the common crab, as Darwin's own grandfather Erasmus had noticed, stationing sentinels when its friends are molting. There were the pelicans forming a wide half circle and paddling toward the shore to entrap fish. There was the house sparrow who "shares any food" and the white-tailed eagles spreading apart high in the sky to get a full view before vocalizing to one another when a meal is spotted. And, of course, there were the greater hordes of mammals:

deer, antelope, elephants, wild donkeys, camels, sheep — for all of whom "mutual aid is the rule". Future ethological studies would not support all of Kropotkin's examples, but the general picture seemed clear to him at the time: Life was a struggle, and in that struggle the fittest did survive. But the answer to the question, "By which arm is this struggle chiefly carried on?" and "Who are the fittest in the struggle?" made abundantly clear that "natural selection continually seeks out ways precisely for avoiding competition". Putting limits on physical struggle, sociability left room for the development of better moral feelings (Kropotkin, 1902).

The message was obvious. "Don't compete! Competition is always injurious to the species, and you have plenty of resources to avoid it", he wrote. "That is the watchword which comes to us from the bush, the forest, the river, the ocean... Therefore combine, practice mutual aid! That is the surest means of giving to each other and to all the greatest safety, the best guarantee of existence and progress, bodily, intellectual, and moral" (Kropotkin, 1902). If capitalism had allowed the industrial 'war' to corrupt man's natural beginnings; if over population and starvation were the necessary evils of progress — Kropotkin was having none of it. Darwin's Malthusian 'bulldog' Huxley had gotten it precisely the wrong way around. Far from having to combat his natural instincts in order to gain a modicum of morality, all man needed to find goodness within was to train his gaze at nature beside him.

Was solace to be found in the evolution of ethics or the ethics of evolution? Huxley put his money on the civilizing process; Kropotkin — who became a famous anarchist — in the call to abandon institutions and laws. Both men saw themselves as walking in Darwin's footsteps, but each had reached an opposite solution. Their public row, heartfelt and public, set the terms of the debate for future generations.

3. Individual versus superorganism

R.A. Fisher was just 23 years old when he delivered a talk to the Eugenics Education Society in London in 1913. "From the moment we grasp, firmly and completely", he projected:

"Darwin's theory of evolution, we begin to realize that we have obtained not merely a description of the past, or an explanation of the present, but a veritable key to the future; and this consideration becomes the more forcibly impressed on us the more thoroughly we apply the doctrine; the more clearly we see that not only the organization and structure of the body, and the cruder physical impulses, but that the whole constitution of our ethical and aesthetic nature, all the refinements of beauty, all the delicacy of our sense of beauty, our moral instincts of obedience and compassion, pity or indignation, our moments of religious awe, or mythical penetration — all have their biological significance, all (from the biological point of view) exist in virtue of their biological significance" (Kohn, 2004).

Huxley had cautioned man to take a good look at Nature and then run, morally, in the opposite direction. Nietzsche, on the other hand, saw that ethics were threatened by evolution only if nature was considered improper (Nietzsche, 1887). Having studied mathematics at university, and devised ways to treat selection formally, Fisher now believed that he could show that there were no grounds for such considerations. In what some took to be a mathematical Mendelian-like appendix to *The Origin*, and many others the single greatest book about evolution after Darwin's, Fisher set out to prove what his spiritual grandfather could only imagine: that natural selection had planted and would continue to water the very seeds of human benevolence.

Darwin's own son, Major Leonard Darwin, provided the funds for 'The genetical theory of natural selection' (Fisher, 1930), and in his patron's spirit, Fisher produced a eugenic tome (Leonard Darwin was the Chairman of the British Eugenics Society). Yes, the very power of natural selection stemmed from the fact that it is "a mechanism for generating an exceedingly high degree of improbability" (Edwards, 2000); rejecting blending inheritance, considering the importance of the evolution of dominance, and offering his somewhat cryptic 'fundamental theorem of natural selection', Fisher showed how. But the last five chapters of the book were devoted to man, for when it came to the flourishing of intelligence as well as kindness — two qualities with an exceedingly high improbability of having been born — it was man alone that might frustrate nature's vision. To Fisher, the demon was wealth, for the more possessions man held, the fewer incentives he had to procreate. Abundant children, after all, meant a thinner sliver of the pie for each; why toil only to see the fruits of your labor waste away in the next generation? Contraception might be an option, but it was only the educated who ever thought to use a condom. Since Fisher was certain that the better quality genes for intelligence and morals alike were clustered in the upper strata of society — for how else would they have gotten there if not through Darwinian competition? — the multiplication of the lower classes presented a major problem: God, he thought, had met entropy by introducing natural selection, but human avarice might frustrate his design.

Luckily, Fisher firmly and unselfconsciously believed, God, in his benevolence, offered the dual gift of responsibility and occasion, arming his creatures with the capacity for free will. The 'fundamental theorem' showed that natural selection had done almost all the work for us, inching man closer to his genetic optimum. Still, Fisher told a BBC Radio audience, "in the language of Genesis, we are living in the sixth day, probably rather early in the morning" (Fisher, 1947). As sure as animals, men could navigate their destiny; nothing was yet determined or prearranged. If only family allowances were provided to propel the desirable — meaning the upper classes — to procreate, mankind could resume its natural course. Far from an instrument of decay, heredity would be its confederate. After all was said and done, for looks and intelligence and morals, genetics was its greatest and only hope.

Reverently reading the nineteenth century German philosopher Nietzsche, the Anglican Fisher followed *Zarathustra* in the quest to create a Superman. For him, the hopes of mankind were all placed in the hands of individuals — the strongest and fittest that our race had produced. But across the Atlantic, among vials of tiny termites, the American ecologist Alfred E. Emerson saw things differently. In the jungles of British Guiana, alongside the great Harvard entomologist William Morton Wheeler, he had learned a natural, rather than pen-and-pencil, lesson: Wheeler was convinced that the colonies of social insects were like a single individual. In fact, there really was no difference, he thought, between sterile workers toiling for the nest and a heart pumping for the good of the body. Far from a random aggregation of individuals, a termite colony was actually a 'superorganism', a system so well integrated that it assumed a life of its own (Wheeler, 1928).

In the room next door to Emerson's at the Department of Ecology at the University of Chicago was Sewall Wright, a statistically oriented geneticist conversant with Fisher's mathematics who was just then developing a theory of gene action. Specific cues, Wright thought, activate specific genes in specific ways — mutating them, or turning them on or off. The genes then activate particular physiological gradients, depending on the chemical cue they receive, directing genetically identical organisms to morph into entirely distinct creatures. That's how development works: the organism, Wright saw, was a 'highly self-regulatory system of reaction', an integrated whole (Wright, 1934).

Emerson liked analogies, and immediately perceived that Wright's model applied to superorganismic termites might just solve Darwin's mystery. Like genes and chemical clues and physiological gradients, termite warriors and toilers and queens each had a particular task. Whether a newborn termite would develop into a soldier or worker depended on the relative complex of castes in the colony — the analog to Wright's chemical cues. Cooperation — and altruism — were a byproduct of specialization of function, and existed because individuals were subordinate to the whole. The key, above all, was 'homeostasis', a term coined and popularized by the physiologist Walter B. Canon referring to the properties of self-control, regulation, and maintenance that ensured the stability of internal environments, like body temperature in mammals (Cannon, 1932). Replacing 'internal environments' with 'population', Emerson wrote: "Just as the cell in the body functions for the benefit of the whole organism, so does the individual organism become subordinate to the whole population" (Emerson, 1946). Homeostasis was the solution to the conflict between part and whole, the regulatory function that balanced the interest of the individual with the good of the group.

When it came to man, the lesson seemed clear. Just a few months after the United States dropped atom bombs on Hiroshima and Nagasaki, Emerson wrote: "It is cooperation or vaporization. It is a struggle for existence by means of the cooperation of all mankind, or extinction through unnatural destructive competition between individuals, classes, races and nations already incorporated into a larger independent whole" (Emerson, 1946). Fisher had put his coin on the individual, Emerson, who never met him but knew his writings well, on the collective. This would be a theme that would come to define the argument over altruism.

4. Group selection under attack

V.C. Wynne-Edwards was a professor at Aberdeen, an Englishman with a Welsh name who had lived half his life in Scotland (Borrello, 2010). Like Kropotkin, he had made expeditions to northern lands, where the harsh elements had driven animals to cooperate. If competition existed in nature, the Arctic taught him, it was directed at the environment, and organisms had developed a myriad of social mechanisms to cope. Wynne-Edwards thought he knew what he was talking about: In 1937, on the Baffin Islands' coastline, he had observed that only between one-third and two-fifths of the fulmars in the breeding colony mated while the rest were pushed into marginal territories and often died (Wynne-Edwards, 1939). What an effective way to prevent overpopulation! Even more striking was the chorus of singing accompanying each breeding season: short of using a calculator, it was the best way for the flock to assess its size and, surveying it resources, reproduce accordingly.

It was an idea, some thought, that contradicted Darwinism. Man might exercise birth control, but birds? In the 1930s, T.H. Huxley's grandson, Julian, had sent his student David Lack to the tropics to study Darwin's finches. What Lack found there was that competition for food was rampant, but that slight differences between geographically isolated groups might reduce it. Each group of finches specializing in a particular food in a particular habitat made for a wonderful mechanism to get out of the others' way; gradually, genetically, the populations become distinct. And yet, as powerful as the force of competition, more powerful — since more fundamental — was the instinct to procreate. Individuals were to maximize their fitness, to sire as many progeny as they possibly could. The idea of altruistic birds passing up an opportunity to do so in service of the greater good seemed absolutely absurd (Lack, 1954).

Unless, of course, natural selection was operating on the group, which was exactly what Wynne-Edwards was arguing (Wynne-Edwards, 1962). When the physiology of the singleton came up against the "viability and survival of the stock or race as a whole", group selection was bound to be the victor and individual reproductive restraint the result. It was just like with fishing: If every fisherman set his net to catch just as many fish as he could, the village folk would quickly find themselves "entering a spiral of diminishing returns". If, however, an agreement was reached over the maximum catch for each, depletion of the villager's food source could be avoided. As with fisherman and their catch, so with fulmars (and red grouse and many other birds) and their environment: To prevent exhausting limited resources, numbers could be regulated by social convention for the benefit of all (Wynne-Edwards, 1959).

George Williams was an American fish specialist. No one hated the idea of group selection more than him, but how to fell it, once and for all? Scouring the literature, Williams found what he was looking for in Fisher. In *The Genetical Theory of Natural Selection*, Fisher had argued that sex ratios in any species should always evolve to an equal number of males and females. The reason was simple: a male born into a species with a female-biased sex ratio is at an advantage, since he has more mating opportunities than a female. If genes determine sex, then in a population with more females, the male-making gene will be favored by natural selection. Eventually, the frequency of that gene will reach a point where the sex ratio of the species is now male-biased, whereupon its alternative, a gene promoting the production of females, will be favored. The dialectic would ensure a symmetry of sexes: the wisdom of selection fashioning males and females one-to-one.

This was a deathblow to group selection, and Williams knew it. For imagine a mother parrot mating and then flying off alone to a deserted island in the middle of the sea. Imagine that she lays 10 eggs, half of which hatch into males, the other half into females, and imagine that her daughters do the same. What happens? The population will grow rather quickly: from 10 in the first generation it will balloon into 50 in the next, 250 in the generation after that, and so on. But what if the mother parrot lays 10 eggs that hatch into 9 females and just 1 male, and her daughters do the same: How then would the population grow? In the first generation, as before, there would be 10 parrots (9 females and 1 male), but in the next generation there would be 90 (81 females and 9 males) and in the one after that 810 (729 females and 81 males). Compared with a 1:1 sex ratio, a female-biased ratio would run much faster: before long the island would be teeming with parrots.

And yet even though siring more females would benefit the growing group, it would reduce the individual fitness of the mother who did so. Imagine two mother parrots arriving on the island — A sticking to the 1:1 sex ration, B to the 9:1. Together they will produce 20 offspring in the first generation: $5 + 9 = 14$ females added to $5 + 1 = 6$ males. When these offspring mate among themselves, while each female continues to lay 10 new eggs, the average male sires 23 children (140 divided by 6) since he fathers the offspring of more than 2 females. The results for A and B are surprising: After three generations, A, who stuck to the slower 1:1 ratio but had more males, will have 165 grand-offspring whereas B, who went for the 9:1, will have only 113.

With the sex ratio there was a conflict between community and individual: If the group counts, 'altruistic' B is the winner, but if the individual is selection's client, 'selfish' A prevails. Wynne-Edwards had shown that in times of dearth it is not always to the advantage of the population to breed as much as possible. But assuming sex ratio was an adaptation, when the times of dearth were over one would expect a switch to a female biased sex ratio. Combing through the literature once again, Williams saw that this never actually happened: not in flies, not in pigs or rabbits, not even in human beings. Over time and across all vagaries, the 1:1 sex ratio remained stable — the very prediction that selection working to maximize individual fitness required (Williams, 1966).

It was a mortal blow to the doctrine of the 'greater good' — the theory of group selection — or so Williams thought.

5. Calculating (un)selfishness

It was the end of 1964 and Bill Hamilton had returned from the Brazilian forests with news of his own: In countless species of ants, thrips, wasps, beetles and mites there are extraordinary sex ratios, major departures from fisher's inevitable one-to-one. Hamilton had found them peeling back bark from capirona and kapok trees, stripping weeds of larvae beneath shallow jungle streams, and munching on figs with internal surprises. *Mellitoba acasta* was an example: a tiny parasitic wasp, its female lays her eggs inside the living pupae of bumblebees. When the eggs hatch they eat their way out of the pupa, but not before each female engages in sex with the sole male, their only brother. After all, it makes sense for their mother to use the confined body of the pupa to lay as many female eggs as possible and only one male to inseminate them: Once they've been fertilized, they can fly away to lay their own eggs in another poor bumblebee pupa, while their exhausted brother and lover, the wingless male, wastes away in the abandoned cocoon (Hamilton, 1967).

That same year Hamilton had published the results of his very lonely doctoral work (Segerstrale, 2013). *The Genetical Evolution of Social Behaviour*, parts I and II, introduced an original idea (Hamilton, 1964). For the better part of the century classical population geneticists had defined 'fitness' as the measure of an organism's reproductive success — the more offspring an organism sired, the greater its fitness. A corollary of this definition was that the persistence of any behavior or gene responsible for it which reduced an organism's fitness would be difficult to explain: in time natural selection should see to its demise. But Hamilton now showed that if fitness were redefined to include the progeny of relatives rather than just one's own, in many

cases the Darwinian difficulty of explaining the evolution of altruistic behavior would simply disappear. The key was adopting a 'gene's eye' point of
view, and the Hymenoptera showed precisely why.

In the ancient order — comprised of sawflies, wasps, bees and ants —
female workers are more closely related to their sisters than to their offspring, due to a genetic quirk in their sex-determining heredity called 'hap-
lodiploidy'. All females are born of eggs fertilized by sperm and are diploid,
whereas all males are born of unfertilized eggs and are haploid. Since female sister workers share all their father's genes and half their mother's, their
coefficient of relationship (a term invented by Sewall Wright) is $r = 0.75$,
whereas since they pass only half of their genes on to the next generation the
relationship to their offspring is only $r = 0.50$. Brothers and sisters, on the
other hand, only share one common ancestor, their mother, and are therefore
only half as related as normal siblings, with $r = 0.25$.

The implications were strangely illuminating. Lazy male drones were not
moral slackers but genetic opportunists: They spent their time seeking to
mate rather than work since their relatedness to offspring would be double the relatedness to any sisters they might oblige. Female workers, too,
were shrewd genetic calculators. Neither shirkers of parenthood nor mind-
less toilers, farming the queen as a 'sister producing machine' (Dawkins,
1976) served their interests more generously than did siring their own brood.
Everything depended on perspective: From the point of view of the gene
trying to make its way into the next generation, it made absolutely no difference in whose body it was being carried. In one stroke, a century after
Darwin, 'inclusive fitness' unveiled the mystery of the ants.

But the Hymenoptera were just an illustration for Hamilton; the principle,
he thought, was a general one. And trudging through pages of algebra, he
came up with an elegant mathematical rule: Every altruistic act — such as a
rabbit thumping its legs to warn its friend of the presence of a predator, or
a howler monkey doing the same with its call — would entail both a fitness
cost to the altruist (the chance that he might attract the predator himself), and
a fitness benefit to the receiver (the chance to live another day and produce
more offspring). What Hamilton showed was that, for every social situation,
if the benefit (B), devalued by the relatedness between the two actors (r), was
greater than the cost (C), genes that play a role in bringing about altruistic
behavior could evolve. The greater the relatedness, the greater the chance
for, and scope of, altruism. Hamilton's Rule, as it became known, $r \cdot B > C$,

formalized what Darwin and Huxley and Fisher (and the English geneticist and biochemist J.B.S. Haldane) had all intuited but never proved: altruism was a family affair.

When George Price encountered Hamilton's Rule he was shocked. A chemist turned engineer turned popular science writer turned computer programmer, Price had traveled from America to England in 1967 with the hopes of solving the mystery of the evolution of the family (Harman, 2010). But was Hamilton saying that nepotistic altruism was the very best nature could muster? Writing to Hamilton rather out of the blue, he received the following response: "With man culture did once, in the form of primitive religions, reinforce socialism, but now what we take to be highest in culture has swung strongly against nepotism and the like. Can this be just a higher hypocrisy induced by the need which civilization creates for genetic diversity — and perhaps even racial diversity — in human groups? I wonder whether this is the field in which you think you see some light" (Hamilton, 1968).

Hamilton's reply suggested an even more pallid picture of morality: The very feeling of repugnance from a limited kin-directed goodness may itself be a trick of the genes to bring about a favored behavior; broadening altruism beyond the family would have the salutary effect of increasing overall genetic diversity, a biological end that would place 'pure' moral sentiment in a rather dark, functional light. Was altruism either true but limited or broad but nothing but a sham? And in any case, whatever the level or scope, were humans ultimately slaves to their true genetic masters?

Price went to work, staying up nights in London's public libraries, searching for answers. And then, rather miraculously he thought, he came up with a short and beautiful covariance equation (Price, 1970); so miraculous, in fact, that he ran to the closest church to become a Christian having concluded that God himself must have planted the equation in his brain to help explain to humanity where kindness comes from. What did the equation say? In effect, and after further development (Price, 1972), what the 'Price equation' showed was that the spread of altruism could be tracked via statistical covariance of the character with fitness rather than calculations of the pathways of relatedness. Relatedness could be one way to share altruistic genes, but it need not be the only way; Hamilton's Rule notwithstanding, altruism depended on association, not family. No less importantly, what the equation showed, too, was that selection can work on different levels of the biological hierarchy simultaneously: on the gene, one the individual, or on the group, or even higher. When the interest of the group trumped the interest of the individual, for example, selection could help bring about altruism. "Have you seen how my formula works for group selection?", Price asked Hamilton over the phone, shortly after sending him the equation (Hamilton, 1996). Hamilton had not yet, in fact, and, despite his discovery of extraordinary sex ratios, still thought the doctrine of 'the greater good' a crude mistake. Before long he would be converted: a 1975 paper showing how the evolution of altruism between relatives is just an instance of group selection rather than an alternative explanation for an 'apparent' altruism was the first application of Price's full covariance equation to an evolutionary problem (Hamilton, 1975).

But if the evolution of a trait like altruism could be modeled in an equation, what this meant for altruism, Price considered with a shudder, was that it was never really what you thought it was. "Scratch and altruist and watch a hypocrite bleed", the biologist and philosopher Michael Ghiselin (1974, p. 247) would write a few years later; after all, whether good for the gene, the organism, or the group, altruism was adaptive — an interested, rather than a selfless, affair. Taking the realization to heart, Price decided to devote himself to London's homeless, as if trying to prove by his radical act of altruism that the human spirit — Huxley-like — might transcend the natural imperative. Before long, having given away everything to those he sought to help, he became a vagabond, living rough on the streets and dwindling to little more than skin and bones. Despite Hamilton's best attempts to save his friend, George Price ultimately put an end to his life in a cold squat in Kentish Town, having failed to figure out, as Aristotle had warned, whether the love we feel for others really stems from the love we feel for ourselves.

6. Open questions

Biological altruism is defined by the result of an action: if a brainless amoeba acts in such a way as to reduce its own fitness while providing a fitness benefit to another, it is an altruist. Psychological altruism, on the other hand, is predicated on intent: If I help an old lady cross the road because I have secret designs to be written into her will, then I am not considered an altruist, even if a truck hits and kills me during the process, and the old lady makes it safely to the other side. And yet, despite the distinction, could there be a connection between altruistic acts in amoeba and altruism in humans? (Sober & Wilson, 1998). Just like the actions of the mind-less amoeba, the brain that allows humans to act selflessly is a product of evolution. More subversively, can we really uphold a hard and fast distinction between human and animal intent, relegating the latter to the realm of biological altruism while reserving psychological altruism exclusively for the former? (de Waal, 2008).

What the eight protagonists of our history teach us, bunched as they are into four dyads, is precisely what is at stake when we consider such propositions. First, as Huxley and Kropotkin battled over bitterly: is morality a product of biology or of culture; an evolutionary legacy or a distinctly human transcendence (or perhaps divine gift), divorced from our biology and in no way dictated by it? Ever since George Price's suicide in 1975, the idea that our natures are what they are, among other things, because we are evolved animals with a history, has gained increasing attention — from ethologists, psychologists, geneticists, philosophers, and mathematical modelers, among others (de Waal, 2006, 2009; Joyce, 2006; Keltner, 2009; Nowak & Highfield, 2011; Boehm, 2012). Still, large swaths of humanity continue to resist such a notion with vigor. The mounting scientific evidence notwithstanding, this is one battle that is not likely to be settled soon.

If the evolutionary history of humanity is indeed relevant to the question of morality, the means by which morality was born become important. Here, secondly, the earlier historical debates continue to resonate: Are altruism and morality better seen as individualist beacons, as Fisher believed, or as collective solutions, as Emerson held? Transposed to the level of mechanism: are they due to group selection, ala Wynne-Edwards, or individual selection, as George Williams maintained? Does selection work at the level of the gene, or at different levels simultaneously, as Hamilton and Price ultimately agreed? Today, the levels-of-selection debate is as fierce as it has ever been, serious attempts to resurrect group selection having been mounted, on the one hand, and inclusive fitness having been attacked, on the other. As ever, this debate seems to wed biological and psychological altruism, with lessons from one invariably coloring our understanding of the other, for better or worse (Wilson, 2012; Harman, 2013).

Perhaps most significant will be the debate about the meaning of a biological understanding of altruism and morality for our ethics. If looking at the history of the altruism and morality debate in biology teaches us anything it is that our attempts to understand these phenomena in nature have been particularly susceptible to insinuations of how we wish to see ourselves in the world. That scientific and social thought intermingle has always been true of man's attempt to make sense of the universe, but it has been conspicuously true when it comes to social behavior. Here, invariably, is's and ought's become inextricably linked, as the tragic fate of George Price attests, perhaps most dramatically. And yet, increasingly, the usual calls to evade Hume's Guillotine, sometimes mistakenly referred to as 'the Naturalistic Fallacy', are now once again being challenged. If in fact there is a naturalistic tale to be told about human morality, then why should not is's and ought's be linked, after all? (on the Is/Ought distinction, see contributions by de Waal, Kitcher, and Blackburn in this issue). To be sure, moral progress is not only or even principally an evolutionary story. Whether kin selection or group selection played their roles, whether genes or ecology proved more important, moral progress is embedded in institutions and in law, has a history, and depends on tradition, negotiation and politics (Kitcher, 2011). Still, is it really beyond the pale to consider that our evolved natures might place certain constraints on how we think and act morally? This will become a pressing issue the more we understand the workings of evolution, especially with reference to the brain (Churchland, 2011; Harman, 2012).

In the *Tractatus*, Ludwig Wittgenstein warns his readers: "Even if all possible scientific questions be answered, the problems of life have still not been touched at all. Of course there is then no question left, and just this is the answer." (Wittgenstein, 1974). Wittgenstein was referring principally to language, to the ability to talk about certain things, and the inability to talk about others. But as we remember Huxley and Kropotkin, Fisher and Emerson, Wynne-Edwards and Williams, Hamilton and Price, and the men and women who came after them and are working today, we do well to consider the Austrian philosopher's notion, applied to the science of altruism and morality. Will a biological approach to ethics uncover for us all that we want to know? What kinds of questions will science usefully consider, and what deliberations — and determinations — will ultimately remain beyond its reach? The history of the altruism–morality debate in biology has provided many insights both about ourselves and about the ways in which we come to know ourselves. How we will re-write this history one hundred years hence remains unknown.

References

Blackburn, S. (2014). Human nature and science: A cautionary essay. — Behaviour 151: 229-244.

Boehm, C. (2012). Moral origins: the evolution of virtue, altruism, and shame. — Basic Books, New York, NY.

Borrello, M.E. (2010). Evolutionary restraints: the contentious history of group selection. — University of Chicago Press, Chicago, IL.

Cannon, W.B. (1932). The wisdom of the body. Revised Edition 1963. — W.W. Norton, New York, NY.

Churchland, P.S. (2011). Braintrust: what neuroscience tells us about morality. — Princeton University Press, Princeton, NJ.

Darwin, C. (1859 [1997]). On the origin of species by means of natural selection. — Wordsworth, Ware.

Darwin, C. (1871). The descent of man and selection in relation to sex. — Princeton University Press, Princeton, NJ, p. 103.

Dawkins, R. (1976). The selfish gene. — Oxford University Press, Oxford.

Desmond, A. (1994). Huxley: the devil's disciple. — Michael Joseph, London.

de Waal, F.B.M. (2006). Primates and philosophers: how morality evolved. — Princeton University Press, Princeton, NJ.

de Waal, F.B.M. (2008). Putting the altruism back into altruism: the evolution of empathy. — Annu. Rev. Psychol. 59: 279-300.

de Waal, F.B.M. (2009). The age of empathy: nature's lessons for a kinder society. — Three Rivers Press, New York, NY.

de Waal, F.B.M. (2014). Natural normativity: the "is" and "ought" of animal behavior. — Behaviour 151: 185-204.

Dugatkin, L.A. (2011). The prince of evolution: Peter Kropotkin's adventures in science and politics. — CreatSpace, Charleston, SC.

Edwards, A.W.F. (2000). The genetical theory of natural selection. — Genetics 154: 1419-1426.

Emerson, A.E. (1946). Biological basis of social cooperation. — Ill. Acad. Sci. Trans. 39: 9-18.

Fisher, R.A. (1930). The genetical theory of natural selection. — Clarendon Press, Oxford.

Fisher, R.A. (1947). The renaissance of Darwinism. — The Listener 37: 1001-1009.

Ghiselin, M. (1974). The economy of nature and the evolution of sex. — University of California Press, Berkeley, CA.

Hamilton, W.D. (1964). The genetical evolution of social behaviour, I and II. — J. Theor. Biol. 7: 1-52.

Hamilton, W.D. (1967). Extraordinary sex ratios. — Science 156: 477-488.

Hamilton, W.D. (1968). Letter to George Price, March 26. — W.D. Hamilton Papers, The British Library, London.

Hamilton, W.D. (1975). Innate social aptitudes in man, an approach from evolutionary genetics. — In: Biological anthropology (Fox, R., ed.). Malaby Press, London, p. 133-153.

Hamilton, W.D. (1996). Narrow roads of gene land. — Spektrum, Oxford.

Harman, O. (2010). The price of altruism. — W.W. Norton, New York, NY.

Harman, O. (2012). Is the naturalistic fallacy dead? (And if so, ought it be?) — J. Hist. Biol. 45: 557-572.

Harman, O. (2013). Shakespeare among the ants. — Stud. Hist. Phil. Biol. Biomed. Sci. 44: 114-118.

Huxley, T.H. (1888). The struggle for existence in human society: a programme — Reprinted in Huxley (1898), Evolution and ethics and other essays. — D. Appleton, New York, NY, p. 197-200.

Joyce, R. (2006). The evolution of morality. — MIT Press, Cambridge, MA.

Keltner, D. (2009). Born to be good: the science of a meaningful life. — W.W. Norton, New York.

Kitcher, P. (2011). The ethical project. — Harvard University Press, Cambridge, MA.

Kitcher, P. (2014). Is a naturalized ethics possible? — Behaviour 151: 245-260.

Kohn, M. (2004). A reason for everything: natural selection and the English imagination. — Faber and Faber, London.

Kropotkin, P. (1902). Mutual aid: a factor in evolution. — Reprint Extending Horizons Books, Boston, MA, 1955.

Lack, D.L. (1954). The natural regulation of animal numbers. — Oxford University Press, Oxford.

Nietzsche, F. (1887 [1996]). On the genealogy of morals, trans. Douglas Smith. — Oxford University Press, Oxford.

Nowak, M. & Highfield, R. (2011). Supercooperators: altruism, evolution, and why we need each other to succeed. — Free Press, New York, NY.

Price, G. (1970). Selection and covariance. — Nature 227: 520-521.

Price, G. (1972). Extension of covariance selection mathematics. — Ann. Hum. Genet. 35: 485-490.

Segerstrale, U. (2013). Nature's oracle: a life of W.D. Hamilton. — Oxford University Press, Oxford.

Sober, E. & Wilson, D.S. (1998). Unto others: the evolution and psychology of unselfish behavior. — Harvard University Press, Cambridge, MA.

Tennyson, A. Lord (1850). In memoriam A. H. H. Available online at: http://www.kalliope.org/en/vaerktoc.pl?vid=tennyson/1850

Todes, D. (1989). Darwin without Malthus: The struggle for existence in Russian evolutionary thought. — Oxford University Press, New York, NY.

Wheeler, W.M. (1928). The social insects. — Harcourt Brace, New York, NY.

Williams, G.C. (1966). Adaptation and natural selection. — Princeton University Press, Princeton, NJ.

Wilson, E.O. (2012). The social conquest of earth. — Liveright, New York, NY.

Wittgenstein, L. (1974). Tractatus logico-philosophicus, trans. D.F. Pears and B.F. McGuinness. — Routledge, London, 6.52.

Wright, S. (1934). Genetics of abnormal growth in the guinea pig. — Cold Spring Harbor Symp. Quant. Biol. 2: 137-147.

Wynne-Edwards, V.C. (1939). Intermittent breeding of the fulmar (*Fulmarus glacialis*) with some general observations on non-breeding in sea-birds. — Proc. Zool. Soc. Lond. A 109: 127-132.

Wynne-Edwards, V.C. (1959). The control of population density through social behaviour: a hypothesis. — Ibis 101: 436-441.

Wynne-Edwards, V.C. (1962). Animal dispersion in relation to social behavior. — Oliver and Boyd, Edinburgh.

[When citing this chapter, refer to Behaviour 151 (2014) 167–183]

The moral consequences of social selection

Christopher Boehm *

Goodall Research Center, Department of Biological Sciences and Anthropology,
University of Southern California, Los Angeles, CA 90089-0371, USA
*Author's e-mail address: cboehm1@msn.com

Accepted 30 September 2013; published online 27 November 2013

Abstract

For half a century explaining human altruism has been a major research focus for scholars in a wide variety of disciplines, yet answers are still sought. Here, paradigms like reciprocal altruism, mutualism, and group selection are set aside, to examine the effects of social selection as an under-explored model. To complement Alexander's reputational-selection model, I introduce group punishment as another type of social selection that could have impacted substantially on the development of today's human nature, and on our potential for behaving altruistically. Capital punishment is a decisive type of social selection, which in our past hunter–gatherer environment was aimed primarily against intimidating, selfish bullies, so it is proposed that moral sanctioning has played a major part in genetically shaping our social and political behaviours. Aggressive suppression of free-riding deviants who bully or deceive has made a cooperatively generous, egalitarian band life efficient for humans, even as it has helped our species to evolve in directions that favour altruism.

Keywords

moral evolution, social sanctioning, hunter–gatherers, chimpanzees, bonobos.

1. Introduction

Today hunter–gatherers may ask mythologically how people could have 'evolved' from a nonmoral animal into a moral one, so raising such issues of provenance goes back at least to the beginning of cultural modernity. This means that the question of 'moral origins' has been on human minds far longer than we have had scholars or ecclesiasts writing about such matters. Unfortunately, since Darwin the term has acquired such a confusingly wide range of meanings that if a given study involves questions of right and wrong, and if it looks to both biology and morality, it is likely to qualify —

for example, Ridley's (1996) *The Origin of Virtue*, which is basically about sociobiological models.

Here, drawing directly from Darwin's (1871) work on sexual selection, I consider altruism's evolution strictly in terms of biocultural evolutionary mechanisms as these work over time. My special interest will be in social selection (e.g., West-Eberhard, 1975, 1983; see also Alexander, 1974, 1987; Simon, 1990; Parker, 1998; Nesse, 2007; Boehm, 2008, 2012b; Hrdy, 2009), defined as genetic selection that is accomplished by the social, as opposed to the natural environment.

We will be focusing on the genetic consequences of behaviours found to-day, by which reproductive success is helped or hindered by how ethically a person behaves in a group context. More specifically, we are interested in the group social control of humans, which has served as a selection force that favors the evolution of altruism, and specifically in the ways that economically-independent mobile hunter–gatherers control their own behaviour so as to avoid costly group disapproval or punishment.

2. How does punitive social selection favor altruism?

Alexander (1987) concluded that altruistic cooperation does not follow Trivers' (1971) very popular, well-balanced, dyadic reciprocal-altruism model. What nomadic foragers actually do, is to assume that if you help someone in dire need in the same band today, then in the future other people, those in whatever band you are living in by then, will help you even though you never helped them. Such indirect reciprocity applies to reproductively important matters like sharing meat or seeing to it that safety nets work, and, because demographically-flexible hunting bands are composed mainly of nonkin (Hill et al., 2011), altruistic contributions are essential.

In everyday hunter–gatherer conversations people's social reputations are a major and favorite subject in gossiping (Wiessner, 2005), and having a good reputation as an altruist can pay off with respect to fitness. Alexander emphasizes marital choices, which favor attractive reputational qualities such as being altruistically generous. Or, in a safety-net context, consider Ecuador's Aché foragers. If a household head becomes sick or injured, more emergency help will be forthcoming for those known to have generously helped others in the past, especially if the cost of helping was high, than for those known to have been well-off but routinely stingy (Gurven, 2004).

Because people favor altruists in choosing marriage or foraging partners, this means that often altruists will be pairing up with other altruists (Wilson & Dugatkin, 1997), and such assortative partnering favors the genes of these more generous cooperators. This is because they will reproductively outcompete other pairs that are less altruistic, and this makes reputational selection quite potent. This model of Alexander's provides a broader conceptualization than costly signaling (Bird et al., 2001), for the reputational signals need not be costly and the personal qualities being weighed by individuals in partner choice can be negative, even though Alexander emphasizes positive qualities.

Here I broaden the scope of social selection by focusing on certain controversial individuals, those prone to act selfishly as bullies, thieves, or cheaters, and by bringing in moralistic group sanctioning (Boehm, 1997). Not only will such free-riders tend to rate poorly when it comes to partner choice, but entire hunting bands will be applying collective social control against them aggressively, in ways that can affect fitness drastically — unless they control themselves. In foraging bands, the ultimate sanction is capital punishment, and the agency of self-control is the conscience.

3. How sanctions led to conscience

Conscience has many meanings, but here I discuss our evolutionary conscience, which Alexander (1987) defined purely in terms of reproductive success. Thus a conscience provides an individual with feedback that helps in staying out of trouble with the group, which means adhering to group rules — except when one will profit by breaking them. In this sense, I will be using the term technically to describe the operation of personal self-assessment and self-control in the face of predictable social reactions that punish deviance (Boehm, 2012b).

Conscience functions have some physical correlates: they are partly localized in the prefrontal cortex (Damasio et al., 1990) and they seem to depend on the paralimbic system (Kiehl, 2008) for critical emotional connections that bring empathy into play. While these brain areas and many of the associated functions are far from unique to humans, responses like blushing socially, feeling shame, and morally internalizing group values such as those which favor generosity seem to be ours alone.

Human capacities for empathy that lead to altruism are associated with taking the perspective of others and being moral (Flack & de Waal, 2000), and hunter–gatherers actively promote such altruism. Today, the paradigms best suited for explaining the resulting generous behaviour appear to be mutualism; social selection; and possibly multilevel or group selection. As a distinctively human type of social selection, group punishment brings in a special dimension, and in fact hunter–gatherers of the 'Late-Pleistocene appropriate' type often act as aggressive, moralistic groups to curb, reform, or eliminate serious deviants (Boehm, 1997). Punishers' costs are very low while the deviant can lose fitness radically, so people acting as punitive groups can be quite impactful in helping (inadvertently) to shape their own gene pools.

4. Deep political background

Three living species are descended from Ancestral *Pan*, and living in social dominance hierarchies is a major feature of our evolutionary past (Wrangham, 1987; Wrangham & Peterson, 1996). If we look for antihierarchical coalitions it is apparent that today humans are prepared to crack down on bullies far more definitively than are chimpanzees or bonobos (Boehm, 2012b), so we have taken our own evolutionary course. This is because with humans sanctioning by groups can be fiercely moralized, and because hunters carry lethal weapons.

As bonobo males compete for rank, basically they team up only with their mothers. However, females do routinely form small coalitions with other females in situations of feeding competition, so they can compete strongly when facing off against individual males (Kano, 1992). Rarely, bonobo coalitions can grow larger and fierce; if a male is seriously bullying a female, he can be severely attacked and wounded (Hohmann & Fruth, 2011). This rare large-group response is reminiscent of group social control in humans.

Chimpanzees form coalitions in a wider variety of contexts, including dyadic male partnerships that routinely try to unseat the alpha (de Waal, 1982), and larger, sometimes mixed coalitions that occasionally eject a high-ranking male from the group or kill him (Boehm, 2012a). In captivity, female coalitions ride herd on the alpha male and prevent unwanted male bullying (de Waal, 1990), while in the wild mostly-male raiding coalitions go after outnumbered strangers in no man's land (e.g., Goodall, 1986), and bisexual coalitions may threaten predators like leopards (Byrne & Byrne, 1988) or pythons (Boehm, 1991).

Similar coalitions are found in humans (Boehm, 1999), but in mobile band societies the earmarks of social hierarchy have been removed to such a substantial degree that anthropologists call them egalitarian. Humans cooperate as entire bands when it comes to safety nets, or meat distributions, or in sanctioning deviants, and people also unite as coalitions against natural predators. Human foragers may also gang up to go raiding against neighbors or rarely to engage in intensive warfare, and we also engage in group rituals. Our reputation for collaboration is deserved, and a generally less recognized type of cooperation is the moralistic group sanctioning I am emphasizing. This cultural activity is crucial if egalitarianism is to stay in place, and also if groups are to cooperate effectively.

In assessing antihierarchical alliances in Ancestral *Pan* as of 5–7 million years ago, the least common denominator has to be chimpanzees and bonobos because they have not "moralized" their collective efforts in aggressively controlling overbearing individuals who offend them. Nor do these apes do more than mitigate the hierarchical tendencies that humans in small bands are able to reduce so drastically. Thus, with morality's help Late-Pleistocene humans achieved a long-term state of political egalitarianism that heavily neutralized alpha power (Boehm, 2012b), and therefore they were able to share coveted large game on an equalized basis (e.g., Wiessner, 1996). Social control was their instrument, and a large brain surely was critical in developing a moral worldview (ethos) that enabled humans to reshape society so *definitively* in favor of individual autonomy.

5. An opportunistic evolutionary conscience

By philosophical common sense we think of the conscience as a set of psychological functions that ennoble our social activities by attuning us to behave ethically, following the norms of our group. However, I have referred to a more neutral scientific view, introduced by Alexander (1987), which simply takes the conscience to be an evolved psychological mechanism that looks to maximize our reproductive success.

Thus a conscience not only tells us to behave ourselves; it advises us to cheat on the system a little — as long as a net personal benefit seems likely. Basically, Alexander talks about a small voice that tells us to do things that assist our fitness, and anyone but a psychopath (Hare, 1993) will experience major conscience functions that hook us up with emotions such as shame or remorse, and can lead both to self-control and to personal reform.

It is our blushing internalized sense of shame, combined with fear of the group's capacity to pressure and punish, that keeps most of us humans reasonably well in line with group mores. In contrast, a chimpanzee or bonobo is ruled simply by fear of external domination or punishment.

6. Capital punishment as a moral indicator

Once humans became moral the intensity of the group punishments they employed could stand as a measure of how deeply a given antisocial behaviour was deeply disturbing to group members, or seen as damaging. Capital punishment was the ultimate measure, and here, setting aside other potent sanctioning measures like ostracism or shaming, we will examine such punishment among mobile foragers in some detail. This involves a survey of 50 pure hunter–gatherer societies, all of a type likely to have predominated during the late Pleistocene.

Table 1 shows the individual actions that provoked a moralistic group killing, and almost half of these societies did report incidences of this. By far

Table 1.
Capital punishment in 50 Pleistocene-type foraging societies (from Boehm, 2012b).

Type of deviance	Specific deviance	Societies reporting
Intimidation of group (21 reports)	Intimidation by malicious sorcery	11
	Repeated murder	5
	Action otherwise as tyrant	3
	Psychotic aggression	2
Sexual transgression (6 reports)	Incest	3
	Adultery	2
	Premarital sex	1
Taboo violation (5 reports)	Endangers group by violating taboo	5
Cunning deviance (2 reports)	Theft	1
	Cheating (meat-sharing context)	1
Miscellaneous (4 reports)	Betray group to outsiders	2
	'Serious' or 'shocking' transgression	2
Deviance unspecified (7 reports)		7
Reports of capital punishment		45
Societies reporting capital punishment		24

the most frequent deviance was intimidating others either through physical or supernatural power.

6.1. Killing of bullies

It is clear that hunter–gatherer populations are likely to produce occasional individuals who simply will not heed the egalitarian moral strictures of their groups to behave with humility (e.g., Lee, 1979), or at least to avoid seriously overbearing behaviour that threatens others physically or supernaturally. A moderate bully can be 'put in his place' nonlethally (Boehm, 2012b) — at least, if his conscience allows him to reform. However, a really serious and determined intimidator will have to be liquidated, and Table 1 shows 21 instances of such executions, with some societies being responsible for more than one instance.

Thus, even in using what amounts to a typical limited modern ethnographic time-sample that basically covers just a century or so, and with ethnographies that by their nature are far from complete, nearly half of these societies report killing such a deviant. If our time-sample were a generous 10 millennia instead of a mere century or so, my prediction is that capital punishment of bullies would be found in every society of mobile hunter–gatherers.

There are six incidences of execution for sex crimes, but it is worth noting that some forager societies take premarital intercourse or adultery quite lightly, while even incest is not considered a really serious offense everywhere. There are also five instances of individuals being killed because they violated serious taboos and thereby endangered the entire group, but the main capital crime is acting the bully.

Being egalitarian, all of these societies dislike bullies intensely, and serious upstarts are widely curbed by bands using both lethal and nonlethal sanctions. Consequently, ever since lethal moralistic sanctioning became effective, at the level of phenotype surely our styles of bullying have been constrained by strong social-selection pressures. These same pressures could have been otherwise modifying our basic political nature in important and consistent ways.

6.2. What about cheating free-riders?

It would appear from Table 1 that deviants who *deceive* have been severely punished much less frequently. However, if we consider such deceptive free-riding in the context of lesser sanctions, such as ostracism or shaming, in a subsample of 10 societies I found that in all of them, as a predatory deceptive practice *theft* is noted as being punishable. To a lesser degree this also was true of cheating and lying, but this held only in about half of this smaller sample (Boehm, 2012b). Thus theft appears to be the main deceptive behavior that brings punishment.

Obviously, capital punishment is potent as a selection agency because it curtails reproduction so directly. The earlier in life it happens the greater the curtailment, and it also will prevent the deceased from supporting his offspring or his offspring's offspring later in life. Ostracism and shaming merely diminish positive social contact and remove deviants from cooperation networks, so as mechanisms of social selection they have less severe effects — but they do take place much more frequently. In combination, with respect to the evolution of altruistic traits all these hostile sanctions seem to be focused primarily on bullying and secondarily on theft, while individuals are left more on their own to defend against face-to-face cheaters simply by avoiding them.

7. A specific timeline

By taking the origins aspect of 'moral origins' literally, we can develop some core diachronic hypotheses about how morality could have evolved. These may be useful in informing future studies in this area, including those by biologists, evolutionary psychologists, zoologists, anthropologists, philosophers, and economists. An obvious starting place is an evolutionary timeline.

7.1. Primitive building blocks 5–7 mya

Ancestrally, we had in place empathetic perspective-taking and self-recognition, along with some ability to form political coalitions that united subordinates against individuals whose applications of power were particularly resented. Also in Ancestral *Pan*, we had a group-living ape that could cooperate politically and in sharing its small game (Boehm, 2012b).

7.2. Early hominims

Briefly, in Africa between 4 and 2 million years ago a variety of small-brained terrestrial apes were evolving, but despite constant debate I believe there is no compelling reason to include any one of the fossils discovered so far in the direct human line of descent. This might apply even to *Homo habilis* (see Klein, 1999).

7.3. A large-game scavenger emerges

1.8 million years ago or more, a tall, angular, and much larger-brained early human evolved, and within a few hundred thousand years this first certain *Homo* was fashioning Acheulian hand axes (Klein, 1999). *Homo erectus* seems to have been power-scavenging large game occasionally and probably sometimes quite actively, by ganging up aggressively to drive big cats off of their large kills (Klein, 1999).

Those occasional large carcasses surely would have been shared — most likely with intimidating alpha males exhibiting some decisive priority. However, the scavenged carcasses often were enormous — and in Africa they would not keep. This meant that once a group had bluffed a big cat away from a recent elephant or rhinoceros kill, there would have been enough meat to nourish the entire scavenging team — even with alphas being free to dominate the meat. This high-quality nourishment provided health benefits and extra energy for everyone on the team. However, where scavenged meat came in smaller packages, the alpha advantage would have become much greater.

7.4. Archaic humans and the advent of large game hunting

In the direct human line *Homo erectus* evolved into the still larger-brained archaic *Homo sapiens*, and active hunting of sizable game began to be more frequent, as with occasional killing of sizable groups of equines at 400 000 BP when wooden hunting javelins were in evidence (see Thieme, 1997). However, it was at 250 000 BP that a critical subsistence transition took place; at that point the archaic humans in Africa were beginning to hunt actively — going after medium-sized ungulates like antelope rather than the enormous land mammals that were occasionally scavenged earlier. The name of the game was now pursuit hunting, and this resulted in a regularized and important place being made for medium-sized game in human subsistence (Stiner, 2002).

I have hypothesized that this development was highly significant for the evolution of social control (see Boehm, 2012b): as meat became more important, its equalized sharing needed to be socially regulated — which meant keeping down individualistic alpha tendencies. Bands amounted to cooperative hunting teams that shared the same fate, and, with these not very large ungulate carcasses, hunting could only be efficient for an entire band if dominant individuals who wanted to use their power to control and basically over-consume meat were curbed. Keep in mind that with alphas unrestrained, a mere antelope would not effectively nourish everyone in a band — it had to be shared out 'fairly', in rather modest portions.

7.5. Anatomically and culturally modern humans

If we look to how contemporary mobile foragers deal with this very fundamental meat-division problem, it is remarkably standardized. They invariably treat a sizable carcass as being community rather than individual property, and they put its distribution in the hands of a neutral party who will not try to politicize the meat division and come out ahead (Boehm, 2004). Among mobile egalitarian foragers today this amounts to a cultural universal, and there is reason to believe that such practices became definitive after serious ungulate hunting phased in.

8. Impact on gene pools

Group sanctioning is an aggressive and deliberately manipulative way of solving social problems, and it became a potent tool in the hands of modern humans — who surely had some predictive understanding of their own social and subsistence systems. As a result, within groups social predators were punitively reformed or eliminated while, more generally, negatively-oriented social selection became a powerful force in the biological evolution of human cooperation.

As archaic humans turned to large-ungulate hunting, very likely such selection intensified with greater crackdowns on bullies, and to a lesser degree on thieves or cheaters. In turn, this would have intensified the evolutionary processes that were making for stronger and more effective conscience functions. Thus, a person who was usually prone to bullying, but was socially sensitive and could control himself, would have won out handily over a similar person with less effective self-control.

On this basis I have proposed that starting about a quarter of a million years ago there may well have been a dramatic increase in the rate of evolution of our moral capacity to internalize rules and judge ourselves by them to see what we can get away with as we try to build useful social reputations. By the time we became culturally modern, presumably our moral evolution was complete in its genetic basics, and if McBrearty & Brooks (2000) are correct we might have reached that status by 100 000 to 150 000 BP, rather than later. Thus a major portion of what I am calling conscience evolution might have taken place rather recently, and in a relatively short span of time. If so, the several applicable types of social selection could help scholars in accounting for such rapidity (Boehm, 2012b), and because of similarities to Darwinian sexual selection (i.e., social decisions are driving the selection process) there could have been 'runaway' effects (Nesse, 2007) to further intensify the rate of selection.

9. Three key behavioural dimensions of our evolved morality

With respect to cooperation, there were specific normative rules that all mobile foraging bands developed and enforced. These rules heavily favored generosity, honesty and humility.

9.1. Generosity

Every forager band of the Pleistocene type we are considering here will strongly approve of and actively promote generosity toward others in the band, and emphatically in a typical mobile band most of these people are not kin. These 'Golden Rule' stipulations exist because group members realize that human propensities to be generous outside the family are not all that strong, they need reinforcement. This preaching cannot only promote cooperation, but help to prevent a band's rank and file from being taken advantage of by predatory free-riders (Boehm, 2008).

Such prodding is needed for human degrees of cooperation to take place, for it is doubtful that in some basic way we are nearly as generous as we are selfish. Perhaps that is more an issue for philosophers than for anthropologists, but my main point is that people would not be regularly promulgating rules for being altruistically generous unless there were some useful potential for such behaviour residing in human nature (Boehm, 2012b). In thinking about such larger issues, I believe that philosophers would do well to follow de Waal (2013) and avoid the pessimistic and superficial portrait of an exclusively selfish and egotistic human nature that earlier sociobiological writers like Dawkins (1976) and Ridley (1996) promulgated so effectively.

This nature involves a mixture of probably rather modest but very important degrees of altruism with a strong, predatory, selfish streak that all too readily promotes predatory social deviance outside the family and sometimes within it. It is a function of group social control to protect people from such predators, and in this context it is clearly the more generous individuals who remain at risk — unless they can consolidate their interests, band together actively, and insist aggressively on keeping their moral communities cooperative, egalitarian, and relatively free of social predation. This means that basically group sanctioning has pitted generous altruists against selfish, predatory free-riders, and an equilibrium has been reached in which the two coexist (Boehm, 2012b).

9.2. Honesty

We humans universally prefer to deal with people we can trust, and we resent thieves and cheaters with their unfair tactics. Trivers' (1971) reciprocal-altruism theory led to hypotheses about dedicated, cheater-detection 'modules' in the human brain (Cosmides et al., 2005), while more realistically an arms race was envisioned between altruistic cheater-detectors and deviously predatory free riders, which involved good guys trying to avoid bad guys. However, group sanctioning provides a major alternative to avoidance, for what Trivers (1971) called 'moralistic aggression' can have strong effects in reducing devious free-riding, and such reduction comes in many contexts (Boehm, 2012b).

Thus, we may reasonably assume that everywhere in the Upper Paleolithic serious bullying was potentially punishable by death, that thieving was universally recognized as a serious dishonesty problem and sanctioned, and that lying and cheating were mainly coped with by personal avoidance but sometimes by shaming, criticism, or other group agencies of social control.

9.3. Humility

Being a capable yet humble man among other capable but humble men is a key for male social success in egalitarian hunting bands, and this would apply to a lesser degree to females. In this context humility means (i) never trying to outdo others in a way that puts them down socially (Fried, 1967); (ii) being careful to avoid any semblance of giving orders when one is in a position of leadership (Boehm, 1993); and (iii) avoiding any other behaviour that is suggestive of bullying or setting oneself up as being superior (Boehm, 1999). The exception seems to be in the area of males competing legitimately for females, where individuals seeking pair-bonds sometimes kill their rivals (Knauft, 1991).

The result is the small political society that was once basic to human life everywhere. Egalitarian relations are insisted upon moralistically by subordinates, who impose serious fitness losses on major transgressors by killing them. As we have seen, they have many other anti-hierarchical tools in their kit; for instance, shaming or ostracizing someone inflicts social pain. Unlike capital punishment, however, these lesser sanctions allow for reform, and possessing an efficient evolutionary conscience makes such reform possible through self-control (Boehm, 2012b).

10. Levels of selection

The focus in this paper has been on a group opinion and on how individuals cope with such opinion, which may be suggestive of a group-selection model's being employed (e.g., Wilson & Wilson, 2007). However, for starters, with respect to reputations it is within a single band (or, with respect to marriage partners, within a local nexus of several bands) that hunter–gatherers compete personally for partnerships in this manner. Thus for the reputational type of social selection the level of evolutionary competition is strictly 'individual'. Furthermore, when entire groups actively apply punishment to make a deviant reform or eliminate him, the level of genetic selection is still individual. The group does serve as a critical social environment, but basically the selection impact falls upon individuals.

One might argue that bands superior in effecting social control will be affording better protection to group members against internal social predators, and that hence such groups should be genetically outcompeting groups that were less well policed and hence less cooperative. I have not made this classical group-selection argument (see Darwin, 1871) because any genetic modeling will be compromised by the facts that hunter–gatherer families change bands so frequently and inter-band marriage is common (Hill et al., 2011; see also Kelly, 1995). My emphasis, here, is on selection effects that are likely to have been robust, and these appear to have come at the level of individuals competing for good reputations, or at the level of individuals competing to stay out of serious trouble.

11. Discussion

Moralistic group sanctioning is so predictable and so collectivized that it may well be the most basic kind of human cooperation we experience outside the family. In spite of the various ancestral precursors, the powerful type of social selection that results is distinctly human, and it has resulted in brain functions and states of feeling that are sufficiently unique that they set us apart as moral beings.

In effectuating our social evolution, I believe the mechanisms of natural selection have been sufficiently complex that a number of theories and models will be needed to fully explain the development of behaviours like morality and altruistic generosity. I also believe that Alexander's positively-oriented vision of social selection (especially, choosing altruists in marriage) has been a major but under-emphasized agency in this respect. Here, I have focused upon the still less recognized punitive side of such selection, and I believe its effects on fitness are likely to have been particularly robust because of capital punishment.

With respect to altruism, the division of labor between reputational and punitive modes of social selection is interesting because reputational social selection has favored altruistic traits directly, while punitive social selection had two major effects that both favored altruism less directly. One was to support a conscience, which makes it possible to internalize Golden-Rule values that reinforce altruism and are universally promoted in human bands. The other was to seriously disadvantage the antisocial free-riders who were altruists' worst enemies as genetic competitors.

At the level of everyday life, it is the human conscience that enables many who are prone to transgress to avoid becoming serious targets of angry groups; thus we regulate our personal behaviour in ways that help us to stay on the positive side of the fitness ledger. Among the Late-Pleistocene foragers who put the finishing touches on our genes, humble, honest, generous reputations assisted reproductive success, while unduly selfish, deceptive, or aggressive behaviours damaged it. This impacted altruists and free-riders alike.

An evolving conscience provided a reputation-sensitive mechanism of self-regulation, which could improve or degrade personal fitness depending on its efficiency. For humanity this was the seat of morality, and having a shame-based conscience to make flexible cost-benefit social calculations has given favored individuals (and possibly favored groups) a special edge. The advantages came through culturally-facilitated altruistic cooperation among nonkinsmen, and they were based on people behaving generously, honestly, humbly, and cooperatively. In an ultimate, evolutionary sense, none of this would have been possible without communities that were prepared to engage in active and sometimes even lethal confrontations with the deviants in their midst.

References

Alexander, R.D. (1974). The evolution of social behavior. — Annu. Rev. Ecol. Syst. 5: 325-384.

Alexander, R.D. (1987). The biology of moral systems. — Aldine de Gruyter, New York, NY.

Bird, R.B., Smith, E.A. & Bird, D.W. (2001). The hunting handicap: costly signaling in human foraging strategies. — Behav. Ecol. Sociobiol. 50: 9-19.

Boehm, C. (1991). Lower-level teleology in biological evolution: decision behaviour and reproductive success in two species. — Cult. Dynam. 4: 115-134.

Boehm, C. (1993). Egalitarian behaviour and reverse dominance hierarchy. — Curr. Anthropol. 34: 227-254.

Boehm, C. (1997). Impact of the human egalitarian syndrome on Darwinian selection mechanics. — Am. Nat. 150: 100-121.

Boehm, C. (1999). Hierarchy in the forest: the evolution of egalitarian behaviour. — Harvard University Press, Cambridge, MA.

Boehm, C. (2004). What makes humans economically distinctive? A three-species evolutionary comparison and historical analysis. — J. Bioeconom. 6: 109-135.

Boehm, C. (2008). Purposive social selection and the evolution of human altruism. — Cross-Cult. Res. 42: 319-352.

Boehm, C. (2012a). Ancestral hierarchy and conflict. — Science 336: 844-847.

Boehm, C. (2012b). Moral origins: the evolution of altruism, virtue, and shame. — Basic Books, New York, NY.

Byrne, R.W. & Byrne, J.M. (1988). Leopard killers of Mahale. — Nat. Hist. 97: 22-26.

Cosmides, L., Tooby, J., Fiddick, L. & Bryant, G.A. (2005). Detecting cheaters. — Trends Cogn. Sci. 9: 505-506.

Damasio, A.R., Tranel, D. & Damasio, H. (1990). Individuals with sociopathic behaviour caused by frontal damage fail to respond automatically to social stimuli. — Behav. Brain Res. 41: 81-94.

Darwin, C. (1871). The descent of man and selection in relation to sex. — John Murray, London.

Dawkins, R. (1976). The selfish gene. — Oxford University Press, New York, NY.

de Waal, F.B.M. (1982). Chimpanzee politics: power and sex among apes. — Harper & Row, New York, NY.

de Waal, F.B.M. (1990). Peacemaking among primates. — Harvard University Press, Cambridge, MA.

de Waal, F.B.M. (2013). The bonobo and the atheist: in search of humanism among the primates. — W.W. Norton, New York, NY.

Flack, J.C. & de Waal, F.B.M. (2000). 'Any animal whatever': Darwinian building blocks of morality in monkeys and apes. — J. Consci. Stud. 7: 1-29.

Fried, M.H. (1967). The evolution of political society. — Random House, New York, NY.

Goodall, J. (1986). The chimpanzees of Gombe: patterns of behaviour. — Belknap Press, Cambridge.

Gurven, M. (2004). To give and to give not: the behavioral ecology of human food transfers. — Behav. Brain Sci. 27: 543-583.

Hare, R. (1993). Without conscience: the disturbing world of the psychopaths among us. — Guilford Press, New York, NY.

Hill, K.R., Walker, R., Bozicevic, M., Eder, J., Headland, T., Hewlett, B., Hurtado, A.M., Marlowe, F., Wiessner, P. & Wood, B. (2011). Coresidence patterns in hunter–gatherer societies show unique human social structure. — Science 331: 1286-1289.

Hohmann, G. & Fruth, B. (2011). Is blood thicker than water? — In: Among African apes: stories and photos from the field (Robbins, M.M. & Boesch, C., eds). University of California Press, Berkeley, CA, p. 61-76.

Hrdy, S.B. (2009). Mothers and others: the evolutionary origins of mutual understanding. — Belknap Press, Cambridge.

Kano, T. (1992). The last ape: pygmy chimpanzee behaviour and ecology. — Stanford University Press, Stanford.

Kelly, R.L. (1995). The foraging spectrum: diversity in hunter–gatherer lifeways. — Smithsonian Institution Press, Washington, DC.

Kiehl, K.A. (2008). Without morals: the cognitive neuroscience of criminal psychopaths. — In: Moral psychology, Vol. 1: the evolution of morality: adaptations and innateness (Sinnott-Armstrong, W., ed.). MIT Press, Cambridge, MA, p. 119-154.

Klein, R.G. (1999). The human career: human biological and cultural origins. — University of Chicago Press, Chicago, IL.

Knauft, B.M. (1991). Violence and sociality in human evolution. — Curr. Anthropol. 32: 391-428.

Lee, R.B. (1979). The !Kung San: men, women, and work in a foraging society. — Cambridge University Press, Cambridge.

McBrearty, S. & Brooks, A. (2000). The revolution that wasn't: a new interpretation of the origin of modern human behaviour. — J. Human Evol. 39: 453-563.

Nesse, R.M. (2007). Runaway social selection for displays of partner value and altruism. — Biol. Theor. 2: 143-155.

Parker, S.T. (1998). A social selection model for the evolution and adaptive significance of self-conscious emotions. — In: Self-awareness: its nature and development (Ferrari, M.D. & Sternberg, R.J., eds). Guildford Press, New York, NY, p. 108-136.

Ridley, M. (1996). The origins of virtue: human instincts and the evolution of cooperation. — Penguin Books, New York, NY.

Simon, H.A. (1990). A mechanism for social selection and successful altruism. — Science 250: 1665-1668.

Stiner, M.C. (2002). Carnivory, coevolution, and the geographic spread of the genus *Homo*. — J. Archaeol. Res. 10: 1-63.

Thieme, H. (1997). Lower Paleolithic hunting spears from Germany. — Nature 385: 807-810.

Trivers, R.L. (1971). The evolution of reciprocal altruism. — Q. Rev. Biol. 46: 35-57.

West-Eberhard, M.J. (1975). The evolution of social behavior by kin selection. — Q. Rev. Biol. 50: 1-33.

West-Eberhard, M.J. (1983). Sexual selection, social competition, and speciation. — Q. Rev. Biol. 58: 155-183.

Wiessner, P. (1996). Leveling the hunter: constraints on the status quest in foraging societies. — In: Food and the status quest: an interdisciplinary perspective (Wiessner, P. & Schiefenhovel, W., eds). Berghahn Books, Oxford, p. 171-191.

Wiessner, P. (2005). Norm enforcement among the Ju/'hoansi bushmen: a case of strong reciprocity? — Hum. Nat. 16: 115-145.

Wilson, D.S. & Dugatkin, L.A. (1997). Group selection and assortative interactions. — Am. Nat. 149: 336-351.

Wilson, D.S. & Wilson, E.O. (2007). Rethinking the theoretical foundation of sociobiology. — Q. Rev. Biol. 82: 327-348.

Wrangham, R.W. (1987). African apes: the significance of African apes for reconstructing social evolution. — In: The evolution of human behaviour: primate models (Kinzey, W.G., ed.). State University of New York Press, Albany, NY, p. 51-71.

Wrangram, R.W. & Peterson, N. (1996). Demonic males: apes and the origins of human violence. — Houghton Mifflin, New York, NY.

Natural normativity: The 'is' and 'ought' of animal behavior

Frans B.M. de Waal *

Living Links, Yerkes National Primate Research Center, Emory University,
Atlanta, GA, USA
* Author's e-mail address: dewaal@emory.edu

Accepted 14 October 2013; published online 27 November 2013

Abstract
The evolution of behavior is sometimes considered irrelevant to the issue of human morality, since it lacks the normative character of morality ('ought'), and consist entirely of descriptions of how things are or came about ('is'). Evolved behavior, including that of other animals, is not entirely devoid of normativity, however. Defining normativity as adherence to an ideal or standard, there is ample evidence that animals treat their social relationships in this manner. In other words, they pursue social values. Here I review evidence that nonhuman primates actively try to preserve harmony within their social network by, e.g., reconciling after conflict, protesting against unequal divisions, and breaking up fights amongst others. In doing so, they correct deviations from an ideal state. They further show emotional self-control and anticipatory conflict resolution in order to prevent such deviations. Recognition of the goal-orientation and normative character of animal social behavior permits us to partially bridge the is/ought divide erected in relation to human moral behavior.

Keywords
morality, normativity, conflict resolution, community concern, fairness, inequity aversion, emotional control.

1. Introduction

One of the most vexing problems facing attempts to ground morality in biology is the so-called is/ought divide. Whether it is a real divide, or not, depends partly on how we phrase the question. Its initial formulator, David Hume (1739), did not in fact see it as a sharp divide. Almost three centuries ago, he asked us to be careful not to assume that we can derive 'ought' from 'is', adding that we should give a reason for trying. Having noticed how often authors move from descriptions of how things are to statements about how things ought to be, he added:

"This change is imperceptible; but is however, of the last consequence. For as this ought, or ought not, expresses some new relation or affirmation, 'tis necessary that it should be observed and explained; and at the same time that a reason should be given; for what seems altogether inconceivable, how this new relation can be a deduction from others, which are entirely different from it" (Hume, 1739, p. 335).

In other words, the way we feel humans ought to behave is not simply a reflection of human nature. Just as one cannot infer traffic rules from the description of a car, one cannot infer moral codes from knowing who or what we are. Hume's point is well taken, but a far cry from the expansion by some later philosophers, who turned his appeal for caution into "Hume's guillotine", claiming an unbridgeable chasm between 'is' and 'ought' (Black, 1970). There exists by no means agreement on this topic, which is why it remains a perennial of philosophical debate, but some have gone so far as to wield this guillotine to kill off any and all attempts, even the most cautious ones, to apply evolutionary logic or neuroscience to human morality. Science cannot tell us how to construe morality, they argue. This may well be true, but science does help explain why certain outcomes are favored over others, hence why morality is the way it is. For one thing, there would be no point in designing moral rules that are impossible to follow, just as there would be no point in making traffic rules that cars cannot obey. This is known as the 'ought implies can' argument. Morality needs to suit the species it is intended for.

'Is' and 'ought' are like the yin and yang of morality. We have both, we need both, they are not the same, yet they cannot be completely disentangled. They complement each other (see also Kitcher, 2014, this issue). Hume (1739) himself ignored the 'guillotine' named after him by stressing how much human nature matters: he saw morality as a product of the emotions, placing empathy (which he called sympathy) at the top of his list. This opinion represented no contradiction on his part, since all that he urged was caution in moving from how we are to how we ought to behave (Baier, 1991). He never said that such a move was prohibited, although he might not have agreed with Singer (1973) for whom the debate about the is/ought divide is a 'triviality' entirely dependent on the definition of morality.

While I concur with many philosophers that it is hard, perhaps impossible, to reason from the level of how things are to how things ought to be, here I will explore whether the divide is equally wide if we leave the conceptual domain and enter that of actual behavioral tendencies and motivations. What if morality is not rationally constructed, but grounded in emotional values, as Hume thought? What if biology is not just on the 'is' side of the equation, but informs us also about the 'ought' side, such as by explaining which values we pursue and for what evolutionary reason? Every organism strives for certain outcomes. Survival is one, reproduction is another, but many organisms also pursue social outcomes that come close to those supported by human morality.

That animal behavior is not free of normativity (defined as the adherence to an ideal or standard) is hardly in need of argument. Take the spider's reaction to a damaged web. If the damage is extensive she will abandon her web, but most of the time she will go into repair mode, bringing the web back to its previous functional state by filling holes or tightening damaged threads by laying new ones (Eberhard, 1972). Similarly, disturbing an ant nest or termite hill leads to immediate repair as does damage to a beaver dam or bird nest. Nature is full of physical structures built by animals guided by a template of how the structure ought to look. This template motivates repair or adjustment as soon as the structure deviates from the ideal. In other words, animals treat these structures in a normative fashion. I am not necessarily thinking here of normative judgment. It is unclear if the animals themselves feel an obligation to behave in a particular way, nor do I assume that every single individual of a large colony has a conception of the whole nest, but it is undeniable that animals collectively or individually pursue goal states.

The question here is whether they do the same with regards to social relations and society at large. Do they seek certain social outcomes and correct or discourage deviations from expectations? Do they take a normative approach to social relationships, and if so, is it guided by the same kind of emotions and values that underlie human morality? Churchland (2011, p. 175) hints at a move from the social emotions to moral values, writing that "basic emotions are Mother Nature's way of orienting us to do what we prudently ought". The question here is whether the same move is recognizable in other species.

2. Social hierarchy and impulse control

The opposite of morality is that we just do 'what we want', the underlying assumption being that what we want is not morally good. This remains a common religious argument against naturalized ethics (Gallagher, 2004). In this view, morality rests on the uniquely human ability to inhibit natural tendencies (Huxley, 1894). For example, Kitcher (2006) labeled chimpanzees 'wantons', defined as creatures vulnerable to whichever impulse strikes them. Somewhere in our evolution we overcame this wantonness, which is what made us human. According to Kitcher (2006, p. 136), this process started with the "awareness that certain forms of projected behavior might have troublesome results".

Myriad animals live with similar knowledge, though, not only when they try to avoid detection by predators or prey through the suppression of sound and movement, but also in the social domain. A dominance hierarchy is one giant system of social inhibitions, which is no doubt what paved the way for human morality, which is also such a system. Impulse control is key to avoid 'troublesome results'. In macaques and other primates low-ranking males vary their behavior dependent on the presence or absence of the alpha male. As soon as alpha turns his back, other males approach females. Putting this principle to the test, low-ranking males refused to approach females so long as the dominant looked on from inside a transparent box, yet as soon as this male was removed, the same males freely copulated with females. These males also took the occasion to perform the typical bouncing displays of high-status males. After such episodes, however, they were excessively nervous upon reunion with the alpha male, greeting him with such wide submissive teeth-baring that the experimenters interpreted their behavior as an implicit recognition that they had violated a social code (Coe & Rosenblum, 1984). Perhaps social rules are not simply obeyed in the presence of dominants and forgotten in their absence, but internalized to some degree. However, when scientists have tried to measure the degree of internalization of human-imposed rules in dogs, by studying their guilty-looking demeanor after violations, they have not found much beyond a direct effect of the owner's behavior on the dog (Vollmer, 1977; Horowitz, 2009).

Not only low-ranking individuals, but also high-ranking ones benefit from impulse control. For example, an alpha male chimpanzee (*Pan troglodytes*) may receive a pointed challenge from a younger male, who throws rocks in his direction or makes an impressive charging display, with all his hair on end. This is a way of testing alpha's nerves. Experienced dominant males totally ignore the din, however, as if they barely notice, thus forcing their challenger to either give up or escalate (de Waal, 1982).

Inhibitions associated with the hierarchy ultimately come about through punishment. After having deprived a large troop of rhesus monkeys (*Macaca mulatta*) of water for three hours, a single water-filled basin was made available. All adults came to drink in hierarchical order, but infants and juveniles drank with the highest-ranking males and mingled with the top matriline, thus ignoring the social hierarchy. Only in the third year of life, through increasing exclusions and punishments, did juveniles begin to learn their place in the overall rank-order and converge with their mother's rank (de Waal, 1993).

Since apes develop more slowly than monkeys, youngsters go virtually unpunished for the first four years of life. They can do nothing wrong, such as using the back of a dominant male as a trampoline, stealing food out of the hands of others, or hitting an older juvenile as hard as they can. One can imagine the shock when a youngster is rejected or punished for the first time. The most dramatic punishments are those of young males who have ventured too closely to a sexually attractive female (de Waal, 1982; Figure 1). Young males need only one or two such lessons. From then on, every adult male can make them jump away from a female by a mere glance or step forward. Youngsters thus learn to control their sexual urges, or at least become more circumspect about acting upon them.

The capacity for impulse control can be experimentally tested in the same way that delayed gratification is being tested in children (Mischel et al., 1972; Logue, 1988). Children are given a marshmallow with the promise that if they do not eat it there will be another one coming. Many children have the capacity to wait for minutes. Similarly, both apes (Beran et al., 1999) and monkeys (Amici et al., 2008) will pass up an immediate reward in favor of a better, delayed one. It has further been shown that chimpanzees, like children, play more with toys in the presence of accumulating rewards suggesting attempts at self-distraction in the face of temptation, allowing the apes to delay gratification for up to 18 min (Evans & Beran, 2007). Other studies have shown that apes can override an immediate drive in favor of future needs, an essential aspect of successful action planning (Osvath & Osvath, 2008). The same intertwinement between emotion and cognition known of humans seems to apply to our close relatives, therefore, including

Figure 1. A young male, about 4 years old, has shown too much interest in one of the estrus females, and is now being punished by an adult male, who has taken his foot in his mouth and swings him around. This will serve as a lesson for the rest of the young male's life about the competitiveness of males around sexually attractive females. Photograph by Frans de Waal.

the deliberate control of emotions. Insofar as such control is mediated by the frontal lobes, it should be pointed out that the popular view that this part of the brain is exceptionally developed in our species is erroneous. The human brain is essentially a linearly scaled-up monkey brain (Herculano-Houzel, 2009; Barton & Venditti, 2013).

3. One-on-one normativity

Morality is defined here as a system of rules that revolves around the two H's of Helping or at least not Hurting fellow human beings. It addresses the well-being of others and often puts community interests before those of the individual. It does not deny self-interest, yet curbs its pursuit so as to promote a cooperative society (de Waal, 1996, 2006). This functional definition sets morality apart from customs and habits, such as eating with knife and fork versus with chopsticks or bare hands. The distinction between moral rules and conventions is already clear in young children (Killen & Rizzo, 2014: this issue). Previously, I have distinguished two levels of moral rules: (a) rules at the one-on-one (dyadic) level of social relationships, and (b) rules at the community level (de Waal, 2013). Table 1 summarizes examples at the one-on-one level.

3.1. Reconciliation

The one-on-one level revolves around the preservation of valuable relationships. One of its most common expressions is conflict resolution, first reported by de Waal & van Roosmalen (1979). A typical example concerns two male chimpanzees who have been chasing each other, barking and screaming, and afterwards rest in a tree (Figure 2). Ten minutes later, one male holds out his hand, begging the other for an embrace. Within seconds, they hug and kiss, and climb down to the ground together to groom. Termed a reconciliation, this process is defined as a friendly contact not long after a conflict between two parties. A kiss is the most typical way for chimpanzees

Table 1.
When individuals seek to preserve harmonious social relationships, they apply one-on-one normativity. Their behavior reflects the value attached to good relations. This table offers four examples: restoration of the dominance hierarchy, relationship repair, negative reactions to inequity, and play resumption. In all cases, primates and other animals actively bring a social relationship back to its original state.

Ideal	Deviation	Repair or correct	Restored
Hierarchy	Disobedience or rank challenge	Punish or reassert dominance	Harmony
Close relationship	Conflict	Reconciliation	Harmony
Cooperation	Unequal rewards	Protest or sharing	Harmony
Relaxed play	Hurt partner	Remedial signals	Harmony

Figure 2. The situation after a protracted, noisy conflict between two adult males at a zoo. The challenged male (left) had fled into the tree, but 10 min later his opponent stretched out a hand. Within seconds, the two males had a physical reunion. Photograph by Frans de Waal.

to reconcile, but bonobos do it with sexual behavior (de Waal, 1987), and stumptail macaques wait until the subordinate presents, then hold its hips in a so-called hold-bottom ritual (de Waal & Ren, 1988). Each species has its own way, yet the basic principle remains the same, which is that former opponents reunite following a fight.

Primatology has long been interested in social relationships so that the idea of relationship repair, implied by the reconciliation label, quickly garnered attention. We now know that about thirty different primate species reconcile after fights, and that reconciliation is not limited to the primates. There is evidence for this mechanism in hyenas, dolphins, wolves, domestic goats, and so on. The reason for reconciliation being so widespread is that it restores relationships that have been damaged by aggression but are nonetheless essential for survival. Since many animals establish cooperative relationships within which conflict occasionally arises, mechanisms of repair are essential. The growing field of animal conflict resolution has been reviewed by de Waal (2000) and Aureli & de Waal (2000).

Most of these studies support the Valuable Relationship Hypothesis, which can be formulated thus: "Reconciliation will occur especially between individuals who stand much to lose if their relationship deteriorates". This hypothesis has also been supported by an elegant experiment that manipulated relationship value by promoting cooperation among monkeys, thus increasing their willingness to reconcile after fights (Cords & Thurnheer, 1993). The above ideas have been formalized in the Relational Model, which places conflict in a social context. Aggression is viewed as one of several options for the resolution of conflicts of interest. Other options are avoidance of the adversary (common in hierarchical and territorial species), and the sharing of resources (common in tolerant species). Weighing the costs and benefits of each option, conflict may escalate to the point of aggression after which there still is the option of undoing its damage by means of reconciliation, which option is favored by parties with overlapping interests (de Waal, 2000; Figure 3). Applying the same standardized methodology as primatol-

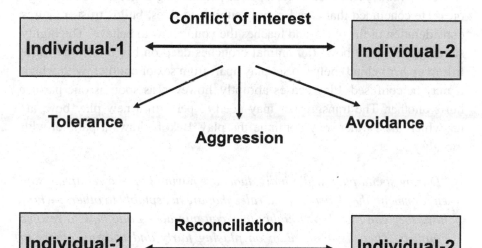

Figure 3. According to the Relational Model, aggressive behavior is one of several ways in which conflicts of interest can be settled. Other possible ways are tolerance (e.g., sharing of resources) and avoidance of confrontation (e.g., by subordinates to dominants). If aggression does occur, it depends on the nature of the social relationship whether or not repair attempts will be made. If there is a strong mutual interest in maintenance of the relationship, reconciliation is likely. Parties negotiate the terms of their relationship by going through cycles of conflict and reconciliation. From de Waal (2000).

ogists to human children, very similar results have been obtained (Verbeek et al., 2000).

3.2. Preventive conflict resolution

That primates guard against the undermining effects of conflict and distress is visible during play. When youngsters are far apart in age, games often get too rough for the younger partner, as when its leg gets twisted or a gnaw turns into a bite. At the slightest peep of distress, its mother will break up the game. Normally, play is entirely silent except for the hoarse panting laughs of apes that resemble human laughter (van Hooff, 1972). Recording hundreds of wrestling bouts, we found that juvenile chimpanzees emit this vocalization especially when the mother of a younger playmate is watching, doing more so in her presence than while alone with the same infant. The vocalizations may function to stave off maternal intervention by reassuring her of the benign nature of the interaction (Flack et al., 2004).

Bekoff (2001) analyzed videotapes of play among dogs, wolves and coyotes. He concluded that canid play is subject to rules, builds trust, requires consideration of the other, and teaches the young how to behave. The highly stereotypical 'play bow' (an animal crouches deep on her forelimbs while lifting up her behind) helps to set play apart from sex or conflict, with which it may be confused. Play ceases abruptly, however, as soon as one partner hurts another. The transgressor may need to perform a new play bow, after which the partner may continue the play. Bekoff draws a parallel with morality:

> "During social play, while individuals are having fun in a relatively safe environment, they learn ground rules that are acceptable to others — how hard they can bite, how roughly they can interact — and how to resolve conflicts. There is a premium on playing fairly and trusting others to do so as well. There are codes of social conduct that regulate what is permissible and what is not permissible, and the existence of these codes might have something to say about the evolution of morality" (Bekoff, 2001, p. 85).

Behavior aimed at the preservation of good relations hints at the great value attached to social harmony. Kummer (1995) offers striking observations of the way hamadryas baboon (*Papio hamadryas*) harem leaders, finding themselves in a fruit tree too small to feed both of their families, will break off their inevitable confrontation by literally running away from each other followed by their respective females and offspring. Chimpanzee males face a similar dilemma. Several of them may sit near a female advertising her fertility with swollen genitals. Rather than competing, the males are actively keeping the peace. Frequently glancing at the female, they spend their day grooming each other. Only when all of them are sufficiently relaxed will one of them try to mate (de Waal, 1982).

The above descriptions are qualitative, but conflict prevention techniques have also been quantified. After an initial suggestion by de Waal (1987) of a grooming peak among captive bonobos right before feeding time, thus preceding potential competition and tension, studies have aimed to measure behavior around food arrival, which at most zoos and institutions occurs at a predictable time of day. Chimpanzees groom more while expecting food, and engage in 'celebrations' marked by high levels of appeasing body contact upon food arrival (de Waal, 1992a; Koyama & Dunbar, 1996). Bonobos, on the other hand, show increased play behavior before food is expected, and high amounts of socio-sexual contact upon its arrival (de Waal, 1987; Palagi et al., 2006). Primates thus anticipate food competition, and actively work to reduce it.

3.3. Striving for fair reward divisions

Negative reactions to skewed reward distributions, also known as inequity aversion (IA), are another part of dyadic relationship maintenance. Cooperative animals need to watch what benefits they obtain relative to their cooperation partners so as not to be taken advantage of. In the absence of equal distribution, mutualistic cooperation might easily become a form of altruism on the part of those who earn less. This outcome problem has been recognized in humans (Fehr & Schmidt, 1999), and is increasingly a theme in animal research (Brosnan, 2011).

Capuchin monkeys are so sensitive to inequity that clumped rewards, which are monopolizable by dominant parties, reduce cooperative tendencies compared to dispersed ones (de Waal & Davis, 2003). Brosnan & de Waal (2003) tested IA in a simple experiment in which two monkeys received either equal rewards for the same task or unequal rewards, such as one monkey receiving cucumber slices and the other grapes, which are far preferred. The authors found that individuals receiving the lesser reward were unaffected if both received the same, yet often refused to perform or accept the reward if their partner received a better deal. Similar results were found in chimpanzees (Brosnan et al., 2005). Experimental replications that did not require a task, however, failed to produce the same results (Braüer et al., 2006; Roma et al., 2006) even in a study on the same monkeys as in the original study (Dindo & de Waal, 2006). Thus, as predicted by an evolutionary account focusing on task performance and cooperation, unequal rewards cause negative reactions only in the context of an effortful task. Finally, van Wolkenten et al. (2007) demonstrated that responses to inequity are truly social in that they cannot be explained as negative reactions to poor rewards while superior ones are visible. Mere visibility had little effect: negative reactions occurred only if the better rewards were actually consumed by a partner.

Similar IA responses have been observed in other species, both primates and nonprimates (Brosnan, 2011; Price & Brosnan, 2012; Range et al., 2012). One restriction, however, is that thus far most studies only concern IA by the individual that receives less. This is known as disadvantageous IA, whereas in advantageous IA subjects respond negatively to receiving a more valuable outcome. Humans show the latter response as well as the former. Brosnan & de Waal (2012) speculate that advantageous IA, which marks a full sense of fairness, occurs when individuals anticipate the negative implications of disadvantageous IA in others. In order to protect the relationship against the eroding effects of tensions when one individual receives less than the other, the one who receives more tries to prevent this by equalizing the outcome. The authors label this a second-order sense of fairness: "In order to prevent conflict within close or beneficial relationships, the advantaged individual will benefit from either protesting or rectifying the situation" (Brosnan & de Waal, 2012, p. 341).

Thus far, there are no signs of second-order fairness in monkeys, such as the capuchin monkeys of the original study. In apes, however, evidence is mounting. The first sign came from a study by Brosnan et al. (2010) on chimpanzees, in which not only partners receiving the lesser reward regularly refused to perform or accept their rewards, but also partners receiving the better reward. In other words, any inequity, not just the disadvantageous kind, was aversive. It made sense, therefore, to test chimpanzees on the Ultimatum Game (UG), which is the gold standard of the human sense of fairness. In the UG, one individual (the Proposer) can split money with another individual (the Respondent). If the Respondent accepts the offer, both players are rewarded, using the proposed split. If the Respondent rejects the offer, however, then neither player is rewarded. People in Western cultures typically offer around 50% of the available amount (Guth, 1995; Camerer & Lowenstein, 2004) as do most other cultures (Henrich et al., 2001). In contrast, a UG study on chimpanzees found them to share the smallest possible amount with the other (Jensen et al., 2006). The methodology of this experiment deviated substantially from the typical human UG, however, and it was unclear if the apes fully understood the task.

To overcome these objections, Proctor et al. (2013) designed a more intuitive UG procedure for both chimpanzees and 3–5-year-old human children. Proposers were presented with a choice of two differently colored tokens that could be exchanged with a human experimenter for food. One color represented an equal reward distribution (3 vs. 3 banana slices), whereas the other represented an unequal distribution favoring the Proposer (5 vs. 1 banana slices). The Proposer would need to hand the token to its partner, the Respondent, sitting behind a mesh panel. Respondents could either accept the token, and return it to the experimenter, or reject it by not returning the token. As in the typical human UG, Proposers thus needed the Respondent's collaboration.

Token choices were compared with choices in the presence of passive Respondents, who lacked any influence. Chimpanzees were sensitive to the contingencies of the game in the same way as humans. If their partner had control, they more often split the rewards equally. In the absence of partner influence, however, they preferred the option that gave themselves the largest proportion of rewards. Since the children behaved similarly, the study suggests that humans and chimpanzees share patterns of proactive decision-making in relation to fair outcomes (Proctor et al., 2013).

4. Community concern

Compared to one-on-one normativity, there are far fewer signs for normativity in nonhuman primates at the community level. This is the level at which human morality may be unique in that we routinely extend our moral reasoning to the society as a whole, speculating what would happen to our community if everyone acted in a particular way. We even extend our value system to interactions that we are not directly involved in. One way in which the moral emotions differ from ordinary ones is "by their disinterestedness, apparent impartiality, and flavor of generality", as Westermarck (1917, p. 238) put it. Typical emotions concern only our personal interests — how we have been treated or how we want to be treated — whereas moral emotions go beyond this. They deal with right and wrong at a more abstract level. It is only when we make judgments of how *anyone* under the circumstances ought to be treated that we speak of moral judgment. To get the same point across, Smith (1759) asked us to imagine how an 'impartial spectator' would judge human behavior.

This is not to say that this level is entirely absent from the behavior of our close relatives. I have previously labeled this level 'community concern' (de Waal, 1996). There exist many examples of impartial policing and mediation that appear to reflect community values. In some species, interventions by the highest-ranking members of the group end fights or at least reduce the severity of aggression. High-ranking male chimpanzees often play this role in fights between females and/or juveniles in their group (de Waal, 1982). For example, if two juveniles are playing and a fight erupts, the alpha male may approach the area of the conflict to stop the fight. By doing so, he reduces the levels of aggression within the group, and also prevents the juvenile fight from escalating by stopping it before the juveniles' mothers intervene and may start fighting amongst themselves.

This pattern of behavior is referred to as the 'control role' (cf. Bernstein & Sharpe, 1966). Detailed descriptions and analyses have been provided by de Waal (1982), together with data showing that males ignore their social ties with the conflict participants while adopting this role. Whereas most individuals support friends and kin, controlling males intervene independently of their usual social preferences (de Waal, 1992b). The ability to put such preferences aside suggests a rudimentary form of justice in the social systems of nonhuman primates. Impartial policing is also known from wild chimpanzees (Boehm, 1994), and a recent study comparing this behavior across various captive groups concluded that it stabilizes social dynamics (von Rohr et al., 2012). An experimental study by Flack et al. (2005), in which key policing individuals were temporarily removed, showed their importance for the maintenance of grooming, play, and other signs of a harmonious society.

One other important method of conflict resolution that has been identified in primate groups is mediation. Mediation occurs when a third party to a conflict becomes the bridge between two former opponents unable to reconcile without external help. It is characterized in the following example (de Waal & van Roosmalen, 1979, p. 62):

"Especially after serious conflicts between two adult males, the two op- ponents sometimes were brought together by an adult female. The female approached one of the males, kissed or touched him or presented towards him and then slowly walked towards the other male if the male followed, he did so very close behind her (often inspecting her genitals) and without looking at the other male. On a few occasions the female looked behind at her follower, and sometimes returned to a male that stayed behind to pull at his arm to make him follow. When the female sat down close to the other male, both males started to groom her and they simply continued when she went off".

Calling such behavior an expression of community concern by no means implies that there are no benefits for the performer. In socially living animals, there exists a great deal of overlap between community-wide and individ- ual interests, and each individual surely has an interest that its community reaches a certain level of harmony and cooperation. The term community concern implies no sacrifice, therefore, and even less selection at the group level. It just states that individuals may advance the interests of their com- munity as a whole, which may well be to their own advantage at the same time that it benefits others.

Finally, prestige and reputation are a critical part of why humans often act on behalf of the community even when they do not directly gain from it. Glimmers of reputation can be seen in the apes. For example, if a major fight gets out of control, bystanders may wake up the alpha male, poking him in the side. Known as the most effective arbitrator, he is urged to step in. Apes also pay attention to how one individual treats another, as in experiments in which they prefer to interact with a human who has displayed a positive attitude to others, such as by sharing food with other apes (Russell et al., 2008; Subiaul et al., 2008; Herrmann et al., 2013). In our own studies, we found that if we let the colony watch two chimpanzees who demonstrate different but equally simple tricks to get rewards, they prefer to follow the higher-status model. Showing a so-called prestige effect, they preferentially imitated prominent members of their community (Horner et al., 2010).

These pieces of evidence suggest that chimpanzees perform actions that benefit the community as a whole and that they may have individual reputations regarding how they treat others or how worthy of imitation they are. This is still a far cry, however, from the human pre-occupation with community standards and the welfare of the whole. It is especially at the level of community concern and reputation building that human moral systems deviate from the normativity found in other primates.

5. Conclusion

In light of the above, the position that biology, including animal behavior, resides entirely on the 'is' side of the is/ought divide is hard to maintain. Obviously, we can describe animal behavior by leaving out any and all references to goals, intentions, and values — just as we can describe human behavior this way — but such descriptions miss an essential aspect. Nonhuman primates, as well as many other animals, strive for specific outcomes. They do so both in relation to physical structures, such as nests and webs, and in relation to social relationships. They actively try to preserve harmony within their social network. They frequently correct deviations from this ideal by, e.g., reconciling after conflict, protesting against unequal divisions, and breaking up fights amongst others. They behave normatively in the sense of correcting, or trying to correct, deviations from an ideal state. They also show emotional self-control and anticipatory conflict resolution in order to prevent such deviations. This makes moving from primate behavior to human moral norms less of a leap than commonly thought.

Differences likely remain, however. Other primates do not seem to extend norms beyond their immediate social environment, and appear unworried about social relationships or situations that they do not directly participate in. They also may not, like humans, feel any obligation to be good, or experience guilt and shame whenever they fail. We do not know if other animals experience such 'ought' feelings. One could argue that their behavior is normative in that it seeks certain outcomes, but that animals manage to do so without normative judgment. They may evaluate social behavior as successful or unsuccessful in furthering their goals, but not in terms of right or wrong. On the other hand, their behavior sometimes suggests a kind of evaluation of past actions, such as when one bonobo bites another and soon thereafter approaches, remembering the exact location of the bite, only to spend half an hour licking the inflicted injury (de Waal, 1989). Given the inaccessibility of animal experience, however, the presence of an internalized normativity remains highly speculative. For the moment, this paper makes the weaker claim, that insofar as the 'ought' of human morality reflects a preference for certain social outcomes over others, similar preferences seem to guide other animals without necessarily implying that they are guided by the same sense of obligation of how they ought to behave as humans.

References

Amici, F., Aureli, F. & Call, J. (2008). Fission-fusion dynamics, behavioral flexibility, and inhibitory control in primates. — Curr. Biol. 18: 1415-1419.

Aureli, F. & de Waal, F.B.M. (2000). Natural conflict resolution. — University of California Press, Berkeley, CA.

Baier, A.C. (1991). A progress of sentiments: reflections on Hume's treatise. — Harvard University Press, Cambridge, MA.

Barton, R.A. & Venditti, C. (2013). Human frontal lobes are not relatively large. — Proc. Natl. Acad. Sci. USA 110: 9001-9006.

Bekoff, M. (2001). Social play behaviour cooperation, fairness, trust, and the evolution of morality. — J. Consc. Studies 8: 81-90.

Beran, M.J., Savage-Rumbaugh, E.S., Pate, J.L. & Rumbaugh, D.M. (1999). Delay of gratification in chimpanzees (*Pan troglodytes*). — Dev. Psychobiol. 34: 119-127.

Bernstein, I. & Sharpe, L. (1966). Social roles in a rhesus monkey group. — Behaviour 26: 91-103.

Black, M. (1970). Margins of precision: essays in logic and language. — Cornell University Press, Ithaca, NY.

Boehm, C. (1994). Pacifying interventions at Arnhem Zoo and Gombe. — In: Chimpanzee cultures (Wrangham, R.W., McGrew, W.C., de Waal, F.B.M. & Heltne, P.G., eds). Harvard University Press, Cambridge, MA, p. 211-226.

Bräuer, J., Call, J. & Tomasello, M. (2006). Are apes really inequity averse? — Proc. Roy. Soc. Lond. B: Biol. Sci. 273: 3123-3128.

Brosnan, S.F. (2011). A hypothesis of the co-evolution of inequity and cooperation. — Front. Decis. Neurosci. 5: 43-55.

Brosnan, S.F. & de Waal, F.B.M. (2003). Monkeys reject unequal pay. — Nature 425: 297-299.

Brosnan, S.F. & de Waal, F.B.M. (2012). Fairness in animals: where to from here? — Soc. Just. Res. 25: 336-351.

Brosnan, S.F., Schiff, H.C. & de Waal, F.B.M. (2005). Tolerance for inequity may increase with social closeness in chimpanzees. — Proc. Roy. Soc. Lond. B: Biol. Sci. 272: 253-258.

Brosnan, S.F., Talbot, C., Ahlgren, M., Lambeth, S.P. & Schapiro, S.J. (2010). Mechanisms underlying responses to inequitable outcomes in chimpanzees, *Pan troglodytes*. — Anim. Behav. 79: 1229-1237.

Camerer, C.F. & Loewenstein, G. (2004). Behavioral economics: past, present, future. — In: Advances in behavioral economics (Camerer, C.F., Loewenstein, G. & Rabin, M., eds). Princeton University Press, Princeton, NJ, p. 3-52.

Churchland, P.S. (2011). Braintrust: what neuroscience tells us about morality. — Princeton University Press, Princeton, NJ.

Coe, C.L. & Rosenblum, L.A. (1984). Male dominance in the bonnet macaque: a malleable relationship. — In: Social cohesion: essays toward a sociophysiological perspective (Barchas, P.R. & Mendoza, S.P., eds). Greenwood, Westport, CT, p. 31-63.

Cords, M. & Thurnheer, S. (1993). Reconciliation with valuable partners by long-tailed macaques. — Ethology 93: 315-325.

de Waal, F.B.M. (1987). Tension regulation and nonreproductive functions of sex in captive bonobos (*Pan paniscus*). — Nat. Geogr. Res. 3: 318-335.

de Waal, F.B.M. (1989). Peacemaking among primates. — Harvard University Press, Cambridge, MA.

de Waal, F.B.M. (1992a). Appeasement, celebration, and food sharing in the two Pan species. — In: Topics in Primatology, Vol. 1, Human origins (Nishida, T., McGrew, W.C., Marler, P. & Pickford, M., eds). University of Tokyo Press, Tokyo, p. 37-50.

de Waal, F.B.M. (1992b). Coalitions as part of reciprocal relations in the Arnhem chimpanzee colony. — In: Coalitions and alliances in humans and other animals. (Harcourt, A. & de Waal, F.B.M., eds). Oxford University Press, Oxford, p. 233-257.

de Waal, F.B.M. (1993). Co-development of dominance relations and affiliative bonds in rhesus monkeys. — In: Juvenile primates: life history, development, and behavior (Pereira, M.E. & Fairbanks, L.A., eds). Oxford University Press, New York, NY, p. 259-270.

de Waal, F.B.M. (1996). Good natured: the origins of right and wrong in humans and other animals. — Harvard University Press, Cambridge, MA.

de Waal, F.B.M. (1998 [orig. 1982]). Chimpanzee politics: power and sex among apes. — Johns Hopkins University Press, Baltimore, MD.

de Waal, F.B.M. (2000). Primates — A natural heritage of conflict resolution. — Science 289: 586-590.

de Waal, F.B.M. (2006). The tower of morality: reply to commentaries. — In: Primates & philosophers: how morality evolved (Macedo, S. & Ober, J., eds). Princeton University Press, Princeton, NJ, p. 161-181.

de Waal, F.B.M. (2013). The bonobo and the atheist: in search of humanism among the primates. — Norton, New York, NY.

de Waal, F.B.M. & Davis, J.M. (2003). Capuchin cognitive ecology: cooperation based on projected returns. — Neuropsychology 41: 221-228.

de Waal, F.B.M. & Ren, R. (1988). Comparison of the reconciliation behavior of stumptail and rhesus macaques. — Ethology 78: 129-142.

de Waal, F.B.M. & van Roosmalen, A. (1979). Reconciliation and consolation among chimpanzees. — Behav. Ecol. Sociobiol. 5: 55-66.

Dindo, M. & de Waal, F.B.M. (2006). Partner effects on food consumption in brown capuchin monkeys. — Am. J. Primatol. 69: 1-9.

Eberhard, W.G. (1972). The web of *Uloborus diversus* (Araneae: Uloboridae). — J. Zool. Lond. 166: 417-465.

Evans, T.A. & Beran, M.J. (2007). Chimpanzees use self-distraction to cope with impulsivity. — Biol. Lett. 3: 599-602.

Fehr, E. & Schmidt, K.M. (1999). A theory of fairness, competition, and cooperation. — Q. J. Econ. 114: 817-868.

Flack, J.C., Jeannotte, L.A. & de Waal, F.B.M. (2004). Play signaling and the perception of social rules by juvenile chimpanzees. — J. Comp. Psychol. 118: 149-159.

Flack, J.C., Krakauer, D.C. & de Waal, F.B.M. (2005). Robustness mechanisms in primate societies: a perturbation study. — Proc. Roy. Soc. Lond. B: Biol. Sci. 272: 1091-1099.

Gallagher, J. (2004). Evolution? No: a conversation with Dr. Ben Carson. — Adventist Review 26 (February).

Guth, W. (1995). On ultimatum bargaining experiments: a personal review. — J. Econ. Behav. Org. 27: 329-344.

Henrich, J., Boyd, R., Bowles, S., Camerer, C., Fehr, E., Gintis, H. & McElreath, R. (2001). In search of *Homo economicus*: behavioral experiments in 15 small-scale societies. — Am. Econ. Rev. 91: 73-78.

Herculano-Houzel, S. (2009). The human brain in numbers: a linearly scaled-up primate brain. — Front. Hum. Neurosci. 3: 1-11.

Herrmann, E., Keupp, S., Hare, B., Vaish, A. & Tomasello, M. (2013). Direct and indirect reputation formation in nonhuman great apes (*Pan paniscus*, *Pan troglodytes*, *Gorilla gorilla*, *Pongo pygmaeus*) and human children (*Homo sapiens*). — J. Comp. Psychol. 127: 63-75.

Horner, V., Proctor, D., Bonnie, K.E., Whiten, A. & de Waal, F.B.M. (2010). Prestige affects cultural learning in chimpanzees. — PLoS-ONE 5: e10625.

Horowitz, A. (2009). Disambiguating the "guilty look": salient prompts to a familiar dog behaviour. — Behav. Process. 81: 447-452.

Hume, D. (1985 [orig. 1739]). A treatise of human nature. — Penguin, Harmondsworth.

Huxley, T.H. (1989 [orig. 1894]). Evolution and ethics. — Princeton University Press, Princeton, NJ.

Jensen, K., Hare, B., Call, J. & Tomasello, M. (2006). What's in it for me? Self-regard precludes altruism and spite in chimpanzees. — Proc. Roy. Soc. Lond. B: Biol. Sci. 273: 1013-1021.

Killen, M. & Rizzo, M. (2014). Morality, intentionality, and intergroup attitudes the origins of morality. — Behaviour 151: 337-359.

Kitcher, P. (2006). Ethics and evolution: how to get here from there. — In: Primates & philosophers: how morality evolved (Macedo, S. & Ober, J., eds). Princeton University Press, Princeton, NJ, p. 120-139.

Kitcher, P. (2014). Is a naturalized ethics possible? — Behaviour 151: 245-260.

Koyama, N.F. & Dunbar, R.I.M. (1996). Anticipation of conflict by chimpanzees. — Primates 37: 79-86.

Kummer, H. (1995). The quest of the sacred baboon. — Princeton University Press, Princeton, NJ.

Logue, A.W. (1988). Research on self-control: an integrating framework. — Behav. Brain Sci. 11: 665-709.

Mischel, W., Ebbesen, E.B. & Raskoff Zeiss, A. (1972). Cognitive and attentional mechanisms in delay of gratification. — J. Person. Soc. Psychol. 21: 204-218.

Osvath, M. & Osvath, H. (2008). Chimpanzee (*Pan troglodytes*) and orangutan (*Pongo abelii*) forethought: self-control and pre-experience in the face of future tool use. — Anim. Cogn. 11: 661-674.

Palagi, E., Paoli, T. & Borgognini Tarli, S. (2006). Short-term benefits of play behavior and conflict prevention in *Pan paniscus*. — Int. J. Primatol. 27: 1257-1270.

Price, S.A. & Brosnan, S.F. (2012). To each according to his need? Variability in the responses to inequity in non-human primates. — Soc. Justice Res. 25: 140-169.

Proctor, D., Williamson, R.A., de Waal, F.B.M. & Brosnan, S.F. (2013). Chimpanzees play the ultimatum game. — Proc. Natl. Acad. Sci. USA 110: 2070-2075.

Range, F., Leitner, K. & Virányi, Z. (2012). The influence of the relationship and motivation on inequity aversion in dogs. — Soc. Justice Res. 25: 170-194.

Roma, P.G., Silberberg, A., Ruggiero, A.M. & Suomi, S.J. (2006). Capuchin monkeys, inequity aversion, and the frustration effect. — J. Comp. Psychol. 120: 67-73.

Russell, Y.I., Call, J. & Dunbar, R.I.M. (2008). Image scoring in great apes. — Behav. Process. 78: 108-111.

Singer, P. (1973). The triviality of the debate over "is-ought" and the definition of "moral". — Am. Philos. Q. 10: 51-56.

Smith, A. (1937 [orig. 1759]). A theory of moral sentiments. — Modern Library, New York, NY.

Subiaul, F., Vonk, J., Barth, J. & Okamoto-Barth, S. (2008). Chimpanzees learn the reputation of strangers by observation. — Anim. Cogn. 11: 611-623.

van Hooff, J.A.R.A.M. (1972). A comparative approach to the phylogeny of laughter and smiling. — In: Non-verbal communication (Hinde, R.A., ed.). Cambridge University Press, Cambridge, p. 209-241.

van Wolkenten, M., Brosnan, S.F. & de Waal, F.B.M. (2007). Inequity responses of monkeys modified by effort. — Proc. Natl. Acad. Sci. USA 104: 18854-18859.

Verbeek, P., Hartup, W.W. & Collins, W.C. (2000). Conflict management in children and adolescents. — In: Natural conflict resolution (Aureli, F. & de Waal, F.B.M., eds). University of California Press, Berkeley, CA, p. 34-53.

Vollmer, P.J. (1977). Do mischievous dogs reveal their "guilt"? — Vet. Med. Small Anim. Clin. 72: 1002-1005.

von Rohr, C.R., Koski, S.E., Burkart, J.M., Caws, C., Fraser, O.N., Ziltener, A. & van Schaik, C.P. (2012). Impartial third-party interventions in captive chimpanzees: a reflection of community concern. — PLoS ONE 7: e32494.

Westermarck, E. (1917 [1908]). The origin and development of the moral ideas, Vol. 2, 2nd edn. — Macmillan, London.

Section 2: Meta-ethics

Introduction

Philosophers have long pondered the relationship between facts and values; between the descriptive and the normative. Are moral principles constrained by empirical knowledge about ourselves and our living conditions? Does the prevailing ecology and local practical knowledge bear upon our judgments of what is right and wrong, what is good and bad? Although conventionally philosophers have argued that the normative world of 'oughts' is unreachable by the practical world of facts, over the last several decades, ethical research has explored new directions. Some philosophers have come to realize that the domain of morality is very complex, and to suspect that the simple 'truisms' about the gulf between facts and values are not true at all. In this vein, Flanagan, Ancell, Martin and Steenbergen open with the crucial question: do the biology and psychology of moral behavior bear upon ethical judgments, and if so, how? Like Aristotle and Dewey, they think of ethics as essentially a practical science, with deep roots in human needs and desires. Like structural engineering, it is fundamentally influenced by design judgments concerning how institutions are best organized.

Blackburn highlights the cautionary approach, taking a long historical perspective, and reminding us of the regrettable tendency to settle for simple answers. Kitcher explains the strengths of John Dewey's understanding of the ethical challenges facing cultures at different historical times and in different geographical conditions. Drumming up absolute rules that supposedly hold for all time in all conditions was not Dewey's approach, and Kitcher shows how the project of understanding human ethical life is ongoing, for conditions are constantly changing, science and technology are ever advancing, and human life is always presenting new challenges. This allows for a modest, sensible kind of pluralism that squares with social reality.

Richard Joyce, finally, takes on the question of whether human ethical behavior is an evolutionary adaptation. His paper tackles the matter of clarifying the question and what would count as evidence either way.

The Editors

[When citing this chapter, refer to Behaviour 151 (2014) 209–228]

Empiricism and normative ethics: What do the biology and the psychology of morality have to do with ethics?

Owen Flanagan [*], Aaron Ancell, Stephen Martin and Gordon Steenbergen

Department of Philosophy, Duke University, 201 West Duke Building,
Box 90743, Durham, NC 27708, USA
[*] Corresponding author's e-mail address: ojf@duke.edu

Accepted 15 September 2013; published online 27 November 2013

Abstract
What do the biology and psychology of morality have to do with normative ethics? Our answer is, a great deal. We argue that normative ethics is an ongoing, ever-evolving research program in what is best conceived as human ecology.

Keywords
empiricism, ethics, eudaimonia, flourishing, moral psychology, moral inference, Hume, naturalism.

> *"Moral science is not something with a separate province. It is physical, biological, and historical knowledge placed in a humane context where it will illuminate and guide the activities of men."* (Dewey, 1922: 204)

1. The question

What do the biology and the psychology of morality have to do with normative ethics? More generally, what does information from the *Geisteswissenschaften*, including the sciences that pertain to our evolutionary history, have to do with how we ought to be and to live, and to the nature of human flourishing? Our answer is a great deal. Ethics concerns the values, virtues, ends, norms, rules, and principles of human personal and social life, all of which are natural phenomena. It casts two lines of inquiry, one into what these features are, and another into what they ought to be. The first line of inquiry is descriptive-genealogical (Flanagan, 1991b, 1996a, 2006;

Flanagan et al., 2007a, b). It aims to identify, describe, explain, and predict the causes, conditions, constituents and effects of the ethical features of human personal and social life, and as such it depends on knowledge about human biology and psychology as well as history, sociology, and anthropology. The second line of inquiry is normative. It aims to say what the features of human personal and social life ought to be — which virtues, values, ends, and practices are good, right, correct, best. Many claim that descriptive-genealogical investigation is irrelevant to normative ethics, that nary the twain shall meet. We defend a modest empiricism: Normative ethics involves the sensible determination of the ends of ethical life and the means to achieve them, where the knowledge on which sensible determination depends is empirical knowledge, construed broadly to include practical local knowledge, accumulated cultural wisdom based on observation, and increasingly the wisdom of the human sciences.[1] Normative ethics is an ongoing, ever-evolving research program in what is best conceived as human ecology (Flanagan, 1996a, b, c, 2002), a project that is, in Dewey's words, 'ineradicably empirical' (1922/1988, p. 295; see also Kitcher, 2011, 2014).

2. The view in a nutshell

Here is the positive view: First, regarding the ends of ethics, the inference to the best explanation is that the *summum bonum* picks out something that everyone sensibly wants — true happiness, flourishing, fulfillment, meaning,

---

[1] Some ethical knowledge is highly contextual and local knowledge, often a kind of know-how that is responsive to what needs to be done here and now in this particular situation among these people. Thus we distinguish between normative ethics as a reflective discipline and normative ethics as the set of practices, often unreflective, that govern moral life on the ground. The discipline is empirical in similar ways to structural engineering (Quine, 1979). Once some wisdom about natural regularities is in place and some experience with building structures starts to accumulate, the discipline of structural engineering emerges. Sensible ends are specified — there is food/work, etc. on the other side of the river — and then there are decisions about whether to build new roads and bridges based on predictions about the likely longevity of the resources on the other side, future demographics, traffic, load, cost-benefit analysis of materials, etc. Then we build. We learn from the experience, and so on. The ordinary on-the ground ethical life of individuals and social groups is more akin to the dynamic lives of the roads and bridges, the water run-off changes and erosion caused by the construction, the wear and tear that the roads and bridges undergo, the effects the surfaces have on cars, and the effects the cars have on them, and so on. All natural, all empirical, all very complicated.

purpose, well-being — something in the vicinity of what Aristotle (1999) called *eudaimonia*.[2] Eudaimonia has subjective components, such as self-respect and self-esteem that are associated with feelings and first-person assessments of the quality of one's life; and objective components that normally involve being warranted in these feelings and judgments, with actually being a good friend, parent, teacher, merchant, fellow citizen. Why the end of having a good human life, eudaimonia, is sought by conscious gregarious social animals with fellow-feeling and reason is not rocket science and its general contours require no defense, although the details are dramatically underspecified until we are in a particular culture, living inside a particular human ecology. Once humans have satisfied the demands of fitness, and resources are above a certain threshold so that there is not or need not be a war of each against each, we turn to flourishing. Commonly, across all the earth, there is a hypertrophic extension of the end of eudaimonia conceived as the desire to live a good human life before we pass away. The *summum bonum* is then expressed as the desire not just for flourishing while alive or the well-being of those who come after us, but for eternal or everlasting flourishing or happiness for oneself — moksha, nirvana, heaven, survival. This too is understandable even if based on a false hope (Flanagan, 2002; Obeyeseker, 2002). In both cases, where eudaimonia is concerned with excellence in this life, or in the cases where there is judgment by God or the impersonal laws of karma, it is decency, reliability, and goodness in this life that matters. Common virtues — compassion, honesty, temperance, courage (Flanagan, 2002; Peterson & Seligman, 2004; Wong, 2006) are best understood as reliable means for achieving the end of living well, of being a good person, or possibly as components or constituents of a good human life. The ubiquity of certain virtues is explained as the outcome of a common human nature

---

[2] Though it is widely thought inside the discipline of ethics that the moral good is the highest good, that the moral life is therefore the best life, and that moral goods always override other goods, many religious thinkers, Kierkegaard most famously, claim that religious goods can override the ethical. In the biblical story of Abraham and Isaac, God demands the 'teleological suspension of the ethical' (1937 [2006]). Contemporary philosophers like Williams (1972), McIntyre (1981 [2007]), and Wolf (1982) have also challenged the overridingness thesis on grounds that there are personal, epistemic, and aesthetic goods that compete with moral goods and which may, all things considered, tip the scales in favor of overriding one's duty. Most people think that at least sometimes the value of new projects or relationships warrant leaving a relationship one has vowed to remain in 'til death do us part'.

faced with common ecological problems across habitats, the most important of which involve various kinds of social dependency.[3]

The idea that normative ethics is an entirely autonomous line of inquiry, forever cut off from empirical scrutiny, grounded instead in some special metaphysics is itself explained by genealogy. The 'project of the enlightenment' (MacIntyre, 1981 [2007]) in the 18th century took upon itself the task of providing secular foundations and rationales for what was once theological ethics. The idea was to provide non-religious reasons for what were otherwise good values, virtues, norms and the like advanced by religion, for example, the Golden Rule. Unfortunately, those engaged in this sensible task also adopted, almost unconsciously, the expectation that ethics could be given what they used to call 'apodictic' grounds, foundations that would give ethics the security of the absolute necessity it (allegedly) had when it was conceived as God's law. And the special foundation sought — a metaphysic of morals — is itself a leftover of the special status given to the set of concerns in terms of which the God(s) of Abraham were supposed to judge the quality of lives. But ethics cannot have such foundations. It can have sensible or reasonable foundations, not necessary ones. We will explain.

3. The ethical domain

The concepts of normative ethics are evaluative and evaluable. Among them are explicit values like freedom, equality, faith, honesty, loyalty, family and friendship, as well as virtues associated with various values. Liberalism values individual freedom, so it prizes and cultivates traits that promote it like independence, responsibility, equality, and respect for law. Classical and contemporary Confucianism by contrast values social order and harmony, and so emphasizes other virtues like elder respect, ritual propriety, graded partiality, and loyalty. Besides explicit values and the virtues associated with them,

---

[3] There is a line of thinking in Philippa Foot (2001) that has both a Christian natural law (Geach, 1956; Anscombe, 1969) and secular form (Thompson, 1995; Foot, 2001) that can be read this way: there are certain natural goods we seek and these are necessary for well-being, and there are some means, for example, the virtues that are necessary for those ends. A modest empiricism recommends weakening the necessity to talk of normal and defensible ends and normal and reliable means, including virtues, for achieving them. And this way of speaking removes most of the distance between Foot's strong cognitivism and Gibbard's (1992) and Blackburn's (1998) expressivism while maintaining the empiricism of both.

there are ultimate ends like happiness, fulfillment, flourishing, heaven and nirvana, each of which defines for a given value system why some values and virtues are to be preferred over others, as well as rules of conduct like the Ten Commandments or the Buddhist Noble Eightfold Path, which typically identify the means to those ends or possibly provide means as well as the recipe for what it is, morality-wise, to live a good human life. Finally, there are principles, like Kant's Categorical Imperative or J.S. Mill's principle of utility, which govern, or are intended to govern morality, and possibly to serve as algorithms to resolve tough choices.

Ethical terms like 'good' and 'bad' emerged originally in ordinary practice among distant ancestors, possibly unreflective ones who were merely conveying likes and dislikes, expressing and endorsing what we, or possibly only I, value. Reflection on the meaning, foundations, and uses of these terms is a recent invention, the product of ethical inquiry and theorizing, which are uniquely human endeavors. However, it would be a mistake to conclude that the ethical domain, conceived as the set of practices we designate as 'moral', is therefore limited to humans. Ethics has no sharp boundaries, except possibly from a point of view internal to an ethical system, which sharply distinguishes it from such nearby, also conventionally defined domains as etiquette and prudence. Dogs learn etiquette — no peeing indoors — as well as prudence — do not cross the street. Are no biting, no chewing, no humping rules matters of dog etiquette, dog prudence or dog proto-morality? Species continuities and adaptive problems across shared environments ensure the appearance of proto-ethical concepts and constructs within the broader field of ethology.

Consider for instance the values of cooperation, fairness and empathy, found to some degree, under specific circumstances, among non-human animals. The degree to which chimpanzees cooperate to coordinate attacks on rival groups is well-documented (de Waal, 1982), but in fact chimpanzees cooperate in other ways as well. Proctor et al. (2013a, b) demonstrate that, like humans, chimpanzees make generally equitable proposals to their partners in a standard economic game, contrary to the predictions of classical utility-based economic models.[4] This suggests that both species value fairness despite immediate economic costs. Chimpanzees also appear to exhibit

---

[4] Proctor et al.'s interpretation is not uncontroversial. See Jensen et al. (2006, 2007).

O. Flanagan et al.

empathy, consoling group members who lose a fight through physical contact (de Waal & Aureli, 1996; De Waal, 2012).[5]

Non-human primates like capuchin monkeys (Brosnan & de Waal, 2003), chimpanzees, and bonobos are not the only species to exhibit proto-ethical behavior. Bates et al. (2008) conclude from thirty-five years of research on elephants that the attribution of empathy best explains several remarkable phenomena: (1) protection of the weak and sick, (2) comforting distressed individuals, (3) 'babysitting', in which a female elephant will temporarily oversee other young, often orphaned individuals, (4) 'retrieval', in which a female elephant will return a young individual to its mother if separated and (5) physically assisting individuals who are stuck or in trouble.

4. Weak and strong theses

Psychology and biology are very much engaged in what was once thought almost exclusively the philosophers' and theologians' domain. Consider a brief catalog of descriptive-genealogical theses. Weak non-imperialistic theories and hypotheses about how science matters to ethics include these: Moral judgments and norms align with variable cultural practices in their surface structure, but at deep levels they pertain to the same domains of life (Shweder et al., 1997, 2002); Moral judgment, like other kinds of judgment, uses a dual process system, sometimes processing quickly and intuitively using reptilian parts of the brain, other times slowly and deliberatively using neo-mammalian parts (Greene, 2007); In virtue of having a common ancestor, the socio-moral aspects of bonobo and chimpanzee life tell us useful things about our pro- and anti-social natures (de Waal, 1982, 1996, 2006, 2008, 2012, 2013); Children cannot lie until they are approximately three years old because they do not have the necessary theory of mind (Gopnik, 2009); Children show pro-social helping behavior in the absence of any external rewards at very young ages (Rheingold, 1982; Warneken & Tomasello, 2006; Dunfield et al., 2011).

Then there are strong theses, some of which contradict each other: Ethics should be taken from the hands of the philosophers and biologized (Wilson, 1975); The moral adequacy of the highest stage of reasoning about

---

[5] Bonobos also exhibit empathy, or proto-empathy, through touching and embracing. See Clay and de Waal (2013).

justice can be inferred from stages of cognitive development (Kant and Rawls win over Mill, Hobbes, Aristotle and the hedonist, in descending order) (Kohlberg, 1973); The neuroscience of decision-making counts against the rule- or principle-based moral theories of Mill and Kant (Churchland, 2011) and in favor of Aristotle's virtue theory (Casebeer, 2003); Science reveals both that humans are generally wired to be 'nice' and that life is meaningless (Rosenberg, 2012); Social psychology reveals that there are no such things as virtue or character traits (Harman, 1999; Doris, 2002); The connectionist computational architecture of the mind favors moral progress over time (Churchland, 1989, 1996a, b, c); There are in human nature several distinct moral foundations or modules (justice, care, disgust, etc.) that are differently tuned up-down/wide-narrow in different cultures and there is no rational way to judge different ways of tuning the modules/foundations (Haidt, 2012); Morality is built upon or out of the neural systems designed to care for kin and this counts against impartial moral theories, such as Peter Singer's (Churchland, 2011).

We could go on. Empirical theorizing about ethics is ubiquitous and increasingly no longer the private turf of ethicists, either philosophical or theological. But there is, the traditional philosopher, the rearguard, will say, this rub. The strong hypotheses above cross the line, they transgress, they make normative inferences from facts; they go from 'is' to 'ought'. And this violates logic, the intellectual division of labor, her majesty the Queen's edicts, and God only knows what else. But this is silly, a red herring. No rational person, scientist or philosopher, tries to derive ought's from is's.

5. From 'is' to 'ought'

Consider four enthymematic arguments. (1) Elephants show fellow-feeling, therefore they should continue to do so. (2) Mammalian care-taking of babies is an adaptation. Human and other mammals should continue caring for their newborns. (3) God disapproves of murder, therefore thou shalt not kill. (4) Humans do not like to be killed, therefore you should not kill them. Not one of the normative conclusions in 1–4 follows deductively. So what? This does not show that the conclusions are false, or even that they are indefensible. It does not even show that they are bad arguments.

But still some will assert that the facts cited have nothing to do with the conclusion and, thus, that the biology and psychology of morality, or any thoroughly descriptive-genealogical account of moral behavior, has nothing to do with what moral behaviors one ought to endorse. And they will do so on the basis of the Humean point that there is a logical difference between facts and values. So, consider Hume. At the end of the first book of his *Treatise on Human Nature* (1739 [2007]), Hume warns against what he finds to be the common practice of deducing ought-conclusions from is-premises. But what does this warning come to exactly? Some have interpreted Hume as having severed any possible connection between facts and values, or at least as having shown that empirical facts are irrelevant to normative ethical inquiry. But in fact, Hume is best read as making a very limited claim, mainly that moral claims are not, strictly speaking, logically deducible, where deduction means derivable or demonstrable using only the laws of deductive logic, from claims about how the world is (Flanagan, 1996a, 2002, 2006; Flanagan et al., 2007a). This does not suggest let alone imply that empirical facts are irrelevant to moral inquiry, only that empirical facts are by themselves insufficient to derive normative conclusions. This is made clear by the fact that Hume himself continues on in the *Treatise* to develop a sophisticated moral theory grounded in empirical observations of human sentiments.

So what does the prohibition on deducing values from facts come to? First, note that in each example above (1–4), we can easily make the argument deductively valid and thus demonstrative by adding a premise to the effect that: It is good for elephants to love their fellow elephants and they should do what is good; or, that it is right to do what God ordains or to refrain from doing what people hate, and so on. The alleged problem then simply devolves into the problem of defending that premise, which itself expresses a normative or value judgment. But why should that worry us? The values expressed are sensible and defensible (Gibbard, 1992; Blackburn, 1998). Elephants, humans, etc. are creatures who have wants, desires, goals and ends. Ethics is one of the spheres of life where gregarious social animals negotiate such things as mutual satisfaction of their aims and interests. And some ends — sharing with family and compatriots, peace, harmony, civility — are just not worth questioning once we are inside 'the ethical project' (Kitcher, 2011).

Furthermore, Hume's warning is one about deducing ought-claims from is-claims, but deduction is not the only sensible kind of reasoning. If non-deductive reasoning were irrational, nearly every body of inferential practices, ordinary common sense ones as well as all the empirical sciences, would be irrational. Hume rules out deduction, but this still leaves us with the

full range of ampliative inferences we regularly use across a wide range of re-
spectable epistemic endeavors. These are the same sorts of inferences we use
anywhere there is a need to project beyond available information: in physical
geodesy, where the earth's shape and density distributions must be inferred
from variations in surface gravity; in medical imaging, where internal structures must be inferred from the scattering of particles, and in the brain's
visual system, where information about the environmental sources of equiv-
ocal retinal projections must be inferred from those projections themselves.
Ethics is no different. We use information about sensible desires, ends, interests, about the circumstances at hand, together with information from past
outcomes (both successes and failures), from our own experience and from
the wisdom codified in the ethical traditions in which we are situated, and
we do the best we can under those circumstances (see Churchland, 1996a, b,
1998; Flanagan, 1996c).

6. The foundations of moral inference

If moral inquiry cannot be, and therefore should not be, the attempt to deduce normative conclusions from factual premises, how, then, should one
figure out what to do and how to live? Patricia Churchland (2009) argues
that decisions about what to do are and ought to be made using what she
calls 'inferences to the best decision'. Inferences to the best decision, like inferences to the best explanation in the sciences, hold a decision or an action
to be justified by virtue of its relative superiority to other alternatives and live
options (see also Thagard & Verbeurgt, 1998). What to do, how to be, and
how to live is determined by comparison to other options and in response to
the situation at hand.

The particular complications of moral decision-making suggest human
beings are likely to make specific sorts of errors in judgment and action,
even by their own lights.[6] The complications are consequences of the sheer

---

[6] Consider an example of how we actually make practical decisions, in this case about how
to distribute scarce resources. In an fMRI study by Hsu et al. (2008), subjects were scanned
as they made decisions about how to distribute a fixed number of meals among children in a
northern Ugandan orphanage. Participants were forced to choose between a less efficient but
equitable distribution (fewer total meals to more children), and an efficient but inequitable
distribution (greater number of meals to fewer children; Hsu et al., 2008). It turns out that
what explains individual decisions to distribute resources in a particular way is, at least in

complexity of calculating what is best to do for oneself and others. Because of this complexity, the brain is liable to either (a) make computational errors when trying to run calculations fully, for example in the attempt to take into account all of the pros and cons of getting married or waging war, or (b) apply shortcuts and rules of thumb to reduce computational complexity, but at the risk of missing some relevant information and making mistakes. For example, stereotyping is an efficient method for deciding how to interact with a member of an outgroup, but it does so at the cost of violating moral rules or principles that one might otherwise endorse. These mistakes can only be remedied if we realize that we make them, understand why we do so, and figure out how to correct ourselves for the better.

Churchland advocates inferences to the best decision to solve problems an individual encounters in the course of moral reasoning, specifically for its sensitivity as a method to biological, psychological, and ecological constraints on choices. The virtues of inferences to the best decision scale up to the theoretical level, where the question is how to conduct normative ethical inquiry, and it is in many respects familiar as a methodological principle within ethics. It is characteristic of what Rawls (1951, 1971) called 'reflective equilibrium' in ethics and political philosophy, a kind of consistency reasoning where one tries to bring one's current judgment about a situation into coherence with one's judgments about similar cases.[7] Some think reflective equilibrium is distinctive in ethics and reveals that ethics does not relate or describe factual truths. It is true that ethics does not relate or describe factual truths. It normally expresses values, recommends and endorses actions, practices, and virtues, and judges actions to be good, bad, right, wrong. But the method of seeking reflective equilibrium is not distinctive of ethics. Self-prediction as well as prediction of the behavior of others turns on assumptions that like cases will be treated (perceived, understood and valued)

---

part, the extent to which they have an aversive affective response to perceptions of inequity. Based on this finding, we can ask questions about how the mechanisms responsible for how we actually make such decisions figure in how we ought to make them. When are perceptions of inequity relevant to how goods ought to be distributed? Is aversion to inequity a reliable mechanism for achieving our ethical interests and concerns regarding distribution? When does it fail?

[7] Wilfrid Sellars makes a similar methodological point about the sciences in that what is characteristic of scientific rationality is the capacity for self-correction by dint of thoroughgoing revisability, the ability "to put *any* claim in jeopardy, though not *all* at once" (Sellars, 1997, p. 79).

in similar ways (Goodman, 1955 [1983]). If I call things that walk and talk like a duck, a duck, then that new thing that walks and talks like a duck is a duck. If this was the way to the water hole yesterday, then I judge it to be the way today. If I invite you to dinner because you are a polite, affable, and interesting guest, then I expect those traits this time. Likewise, if you and I believe that all humans deserve equal treatment but find ourselves stumped or resistant to treating a person of some unfamiliar ethnicity *e* as an equal, then we will have an ethical cramp, which will be resolved either by thinking of some reason why persons of type-*e* are not persons, or why they are persons, but for some other non-arbitrary reason, do not deserve equal treatment, or by judging them to be persons and abiding our principles and extending equal treatment despite our feelings (which we can now work to dissipate).

7. Impartial ethics

What can be said by appeal to broadly empirical, sensible decision procedures like abductive reasoning and methodologies like reflective equilibrium, about how normative ethics ought to be conducted, and which ends and goods humans ought to pursue?

Empirical findings about and reflection on our biological or first nature can tell us about the biological roots of the goods we tend to seek and possibly about reliable means to achieve these goods in their original evolutionary ecologies (many of which antedate hominid evolution). But it would be a mistake to think that investigation into first nature can tell us by itself what the rationally defensible goods and practices ought to be for highly cultured beings living in worlds radically different from the worlds of our ancestors. Consider Patricia Smith Churchland's (2011) critique of impartial ethical theories, for example, Peter Singer's consequentialism (Singer, 1972). On Churchland's view, human ethical behavior is rooted proximately in the neurobiological mechanisms for care of offspring, which manifests itself as other-regarding concern for members of one's immediate family and clan, which is then extended more widely. But there are limits to how far one can extend this other-regarding concern. Since the neurobiological mechanisms of caring were 'designed' to be partial, complete impartiality seems beyond our reach. So, the demands of Singer's impartial consequentialism fall to considerations of evolutionary and psychobiological realism.

But an inclination's basis in our first nature does not alone explain why it cannot, if it cannot, be overridden upon reflection. There are many things that first nature inclines us to do that we think we ought not, as thoughtful, reflective, cultured beings, do. Many — from Jesus to Buddha to John Stuart Mill — have suggested that human flourishing is best served when we consider the interests of everyone as having an equal claim, despite our biological tendency to direct other-regarding concern narrowly or inequitably.[8] Whether it is, as a matter of fact, psychologically unrealistic to suppose that human flourishing is best served when we give impartial consideration to everyone's interests is, indeed, an empirical question. However, it is just not a question answered by facts about first nature alone.

8. The metaphysics of morals is a bad idea

The assumption of a developmental component to human nature, identified by Aristotle (1999) and called our second nature by Burnyeat (1980), is particularly crucial to moral psychology because the degree to which humans are receptive to cultural norms determines the possibilities or constraint space for moral education, including moral self-cultivation, which is the presumed final, application stage of any putative ethical theory. Focusing too much on our first nature obscures the importance of culture to what it is that makes for a good and fulfilling life, and it neglects the role of the social sciences in contributing to normative ethics.

As an example of how conditions and components of well-being and goodness depend not only on our first nature, but on the culture into which we are born, consider the value placed on sacredness in the Middle East. Ginges et al. (2007) conducted experiments with Palestinian and Israeli participants to investigate the role of sacredness in the ongoing conflict in Israel. They found that violent opposition to compromise actually increased when a pro-

---

[8] A lot depends here on whether the principle is interpreted as recommending a heuristic where there is equal *consideration* of interests before action or whether the principle recommends or demands actually acting upon the equal interests of all. Jainism and Buddhism are traditions that recommend something like impartial consideration of the good of all sentient beings without anything like the incredible demand that these interests actually be taken into account practically. This of course invites the response that talk and thought are cheap. P.S. Churchland is focused on action, action-guiding-ness; but wide, even impartial 'consideration' might have good practical effects in certain contexts, in future worlds, and among future persons even if it does not right now, in this situation.

posed peace plan involved material compensation for giving up something considered sacred. However, opposition to compromise decreased when a proposed peace plan involved both sides giving up something sacred. This illustrates how empirical facts can help us determine the best means of reaching our normative ends. In this case, it provides facts that help us determine how to resolve a longstanding violent political conflict. The results of Ginges et al. (2007) also suggest that promoting well-being requires careful attention not just to what material resources people have, but also to the things people value that material resources cannot buy or compensate them for. If people value some things as sacred, then protecting those things may do more to promote those people's well-being than would simply giving them more material resources.

As the skeptic will notice, this does not show that the empirical sciences contribute to the concerns and interests humans ought to have. At best it shows that the empirical sciences are instrumentally relevant to normative ethics, on grounds that normative ethics is concerned with the means to flourish within biological and ecological constraints, and biological and ecological constraints are matters of empirical fact. But if the empirical sciences can contribute knowledge of the causes and conditions of human interests and concerns, and if the interests and concerns of the good life are the ends that human beings ought to have, then the empirical sciences can contribute to the determination of human interests and concerns, in addition to the means we ought to take to achieve them.

The skeptic will continue to ask how one is to justify ethical attention to well-being or flourishing or even just being good. But seeking justification for attention to these things, while not conceptually incoherent, or vulnerable to a knockdown argument, is nonetheless practically misguided (Kitcher, 2011). It is just as misguided as questioning astronomy's attention to the skies. The basic assumption that astronomy should describe, predict, and explain the movements of celestial bodies has not been abandoned despite the gradual broadening of its evidential basis to include information previously thought to be irrelevant to it, like terrestrial physics. Similarly, the basic assumption, one confirmed by empirical investigation, that humans seek meaning, fulfillment, and flourishing beyond their own biological needs, is sufficient to justify attention to the ends, causes and conditions of well-being.

9. Jonathan Haidt's objection to ethical rationality

To deny that empirical inquiry can contribute to ethics in the way we have described has the disastrous consequence of encouraging the belief that those who offer descriptive-genealogical accounts are limited to investigating and evaluating the means to achieving antecedently established ethical ends. Consider recent work by Jonathan Haidt as an example. Haidt (2012) proposes that there are six 'foundations of intuitive ethics'. These foundations are innate psychological mechanisms (Haidt & Joseph, 2004, 2007) which generate moral intuitions in the domains of: (1) care/harm; (2) fairness/cheating; (3) liberty/oppression; (4) loyalty/betrayal; (5) authority/subversion; and (6) sanctity/degradation.[9] The relative sensitivity of these foundations varies between individuals, which is supposed to explain much of the observed diversity in moral opinions. Examples abound. Some people have a very sensitive loyalty/betrayal foundation which makes them more likely to condemn actions like flag burning than people in whom that foundation is less sensitive (Koleva et al., 2012).

This sort of variability prompts us to ask whether some ways of weighting the foundations are better than others. While Haidt does call for tolerance of those whose configuration of foundations differs from our own, and in that regard makes some normative claims (e.g., that harmony is better than conflict and disharmony), he resists arguing that any particular developmental trajectory or configuration of moral foundations is morally better than any other (Haidt, 2012). Sometimes, he outright denies that there are any reasons one can give that favor one moral configuration over another. Indeed, Haidt claims that most philosophers are confused about this. Moral debates are resolved, if they are, in the way philosophers like Thrasymachus, and at times, Nietzsche and Foucault, have suggested: by power, either raw political power as when might makes right, or by the power of seemingly benign socializa-

---

[9] Like other phenotypes, Haidt's moral foundations are taken to be distributed normally across human populations. They are also informationally encapsulated to a point, meaning that a given individual may be highly sensitive to violations triggering one foundation, while remaining insensitive to those triggering another. For example, Haidt (2012) and Graham et al. (2009) suggest that for American liberals, the care/harm and fairness/cheating foundations are far more sensitive than the other four foundations, whereas for American conservatives, all six foundations are weighted more or less equally.

tion.[10] You can fight for your view or brainwash people into believing it, but you cannot say anything rational on its behalf.

Is this view credible? Consider the role of disgust in moral judgment (Haidt et al., 1993; Wheatley & Haidt, 2005). Haidt's claim is that there is nothing rational to say about various extensions of the initial disgust foundation. But there is. Disgust to contaminants makes sense because these things really are dangerous. Evolutionarily, the disgust response system is designed to generate an adverse reaction, repulsion, to certain potentially dangerous contaminants, e.g., excrement, germs, pollutants and poisons. But it is extended in some communities to ground judgments that, for example, homosexuality is wrong because it is disgusting or that members of certain races are slime. But there is a difference: homosexuals and members of different races may disgust you and your people but they are not in fact dangerous.

The point here is a familiar one: Even if disgust is an adaptation in the original evolutionary environment, it does not follow either that it is an adaptation in all current environments or that its extensions are adaptive, in the non-biological senses of being well-suited to contemporary life or sensible or rational as an extended phenotype. The exploitation of any evolved psychobiological trait outside of the environment in which it was selected to operate must be critically assessed. This is what ethics is for.[11,12]

To provide a positive example of how such assessment works, consider the extension of the compassion/care foundation and the justice/harm foundation beyond kith and kin and our local communities. Unlike the extensions of disgust to people and practices that are not in fact contaminating, these extensions do make sense because over world-historical time we have come to

---

[10] Haidt calls this process 'Glauconian' from the character Glaucon in Plato's *Republic*. He should call it 'Thrasymachean' since Thrasymachus, the sophist, defends the view that justice (maybe all of morality) is whatever those in power say it is. Glaucon, Plato's brother, explores the idea without endorsing it.

[11] Several philosophers have made persuasive arguments against using disgust as a basis for moral judgment (see, for example, Knapp, 2003; Nussbaum, 2004; Kelly, 2011). Others have tried to defend a legitimate normative role for disgust (Plakias, 2013).

[12] When Quine suggested that epistemology be naturalized (1969) there was an outcry that this spelled doom for normative epistemology. But this was an over-reaction. We continue to judge the practices of logic, statistics, and probability theory in terms of whether they lead to warranted belief, good predictions, and so on (Flanagan, 2006).

understand that our fate is a shared one, and also that some of the very same reasons that make us care so deeply for loved ones are true of others — they have interests, they suffer, and so on. We care sensibly about consistency, and so for good and defensible reasons we extend or tune up the original cognitive, affective, conative settings to extend more widely.

10. Conclusion: A modest empiricism

There is another unfortunate consequence of denying the contribution of the sciences, of human ecology broadly conceived, to normative ethics. It suggests that special, esoteric expertise is necessary for saying anything serious or meaningful about what ought to be done or how to be and to live, and thus that moral inquiry calls for and somehow involves resources that are not empirical. At a practical level the skeptic does more harm than good by leaving ethics to those much more likely to be under delusions about the sources of ethical normativity.

We should not conclude from the relevance of the biology, psychology and the other human sciences to normative ethics that the study of morality is for the sciences alone. Ethics, by its very nature, is within public domain. Much of the negotiation of how to be good and to live well involves local knowledge, fine-grained attention to the particularities of the self, one's loved ones, and one's community. The study of well-being and the good life (Flanagan, 2007) ought to welcome contributions from any field of inquiry that endorses the idea that ethics is an empirical inquiry into sensible and humane ends for living well and sensible and sensitive means to achieve them.

References

Anscombe, G.E.M. (1969/1981). On promising and its justice. — Collected Philosophical Papers. University of Minnesota Press, Minneapolis, MN, p. 10-21.

Aristotle (1999). Nicomachean ethics (Irwin, T., ed.). — Hackett, Indianapolis, IN.

Bates, L.A., Lee, P.C., Njiraini, N., Poole, J.H., Sayialel, K., Sayialel, S., Moss, C.J. & Byrne, R.W. (2008). Do elephants show empathy? — J. Conscious. Stud. 15: 204-225.

Blackburn, S. (1998). Ruling passions. — Clarendon Press, Oxford.

Brosnan, S. & de Waal, F. (2003). Monkey's reject unequal pay. — Nature 425: 297-299.

Burnyeat, M.F. (1980). Aristotle on learning to be good. — In: Essays on Aristotle's ethics (Rorty, A., ed.). University of California Press, Berkeley, CA, p. 69-92.

Churchland, P.M. (1989). A neurocomputational perspective: the nature of mind and the structure of science. — MIT Press, Cambridge, MA, p. 192-215.

Churchland, P.M. (1996a). Flanagan on moral knowledge. — In: The Churchlands and their critics (McCauley, M., ed.). Blackwell, Cambridge.

Churchland, P.M. (1996b). The neural representation of the social world. — In: Mind and morals (May, L., Freidman, M. & Clark, A., eds). MIT Press, Cambridge, p. 91-108.

Churchland, P.M. (1998). Toward a cognitive neurobiology of the moral virtues: moral reasoning. — Topoi 17: 83-96.

Churchland, P.S. (2009). Inference to the best decision. — In: The Oxford handbook of philosophy and neuroscience (Bickle, J., ed.). Oxford University Press, Oxford, p. 419-430.

Churchland, P.S. (2011). Braintrust: what neuroscience tells us about morality. — Princeton University Press, Princeton, NJ.

Clay, Z. & de Waal, F.B.M. (2013). Bonobos respond to distress in others: consolation across the age spectrum. — PLoS One 8: e55206.

de Waal, F. (1996). Good natured: the origins of right and wrong in humans and other animals. — Harvard University Press, Cambridge, MA.

de Waal, F. (2006). Primates and philosophers: how morality evolved. — Princeton University Press, Princeton, NJ.

de Waal, F. (2008). Putting the altruism back into altruism: the evolution of empathy. — Annu. Rev. Psychol. 59: 279-300.

de Waal, F. (2012). The antiquity of empathy. — Science 336: 874-876.

de Waal, F. (2013). The bonobo and the atheist: in search of humanism among primates. — W.W. Norton, New York, NY.

de Waal, F.B.M. (1982). Chimpanzee politics. — Jonathan Cape, London.

de Waal, F.B.M. & Aureli, F. (1996). Consolation, reconciliation, and a possible cognitive difference between macaques and chimpanzees. — In: Reaching into thought: the minds of the great apes (Russon, A.E., Bard, K.A. & Taylor Parker, S., eds). Cambridge University Press, Cambridge, p. 80-110.

Dewey, J. (1922). Human nature and conduct: an introduction to social psychology. — Henry Holt, New York, NY.

Doris, J.M. (2002). Lack of character: personality and moral behavior. — Cambridge University Press, Cambridge.

Dunfield, K., Kuhlmeier, V.A., O'Connell, L. & Kelley, E. (2011). Examining the diversity of prosocial behavior: helping, sharing, and comforting in infancy. — Infancy 16: 227-247.

Flanagan, O. (1991b). Varieties of moral personality: ethics and psychological realism. — Harvard University Press, Cambridge.

Flanagan, O. (1996a). Ethics naturalized: ethics as human ecology. — In: Mind and morals: essays on cognitive science and ethics (May, L., Friedman, M. & Clark, A., eds). MIT Press, Cambridge, MA, p. 19-44.

Flanagan, O. (1996b). Self expressions: mind, morals and the meaning of life. — Oxford University Press, Oxford.

Flanagan, O. (1996c). The moral network. — In: The Churchlands and their critics (McCauley, R., ed.). Wiley-Blackwell, Hoboken, NJ, p. 192-216.

Flanagan, O. (2002). The problem of the soul: two visions of mind and how to reconcile them. — Basic Books, New York, NY.

Flanagan, O. (2006). Varieties of naturalism: the many meanings of naturalism. — In: The Oxford handbook of religion and science (Clayton, P. & Simpson, Z., eds). Oxford University Press, Oxford, p. 430-452.

Flanagan, O. (2007). The really hard problem: meaning in a material world. — MIT Press, Cambridge, MA.

Flanagan, O., Sarkissian, H. & Wong, D. (2007a). Naturalizing ethics. — In: Moral psychology, Vol. 1: the evolution of morality: adaptations and innateness (Sinnott-Armstrong, W., ed.). MIT Press, Cambridge, MA, p. 1-25.

Flanagan, O., Sarkissian, H. & Wong, D. (2007b). What is the nature of morality: a response to Casebeer, Railton and Ruse. — In: Moral psychology, Vol. 1: the evolution of morality: adaptations and innateness (Sinnott-Armstrong, W., ed.). MIT Press, Cambridge, MA, p. 45-52.

Foot, P. (2001). Natural goodness. — Clarendon Press, Oxford.

Geach, P. (1956). Good and evil. — Analysis 17: 35-42.

Gibbard, A. (1992). Wise choices, apt feelings: a theory of normative judgment. — Harvard University Press, Cambridge, MA.

Ginges, J., Atran, S., Medin, D. & Shikaki, K. (2007). Sacred bounds on rational resolution of violent political conflict. — Proc. Natl. Acad. Sci. USA 104: 7357-7360.

Goodman, N. (1955/1983). Fact, fiction, and forecast. — Harvard University Press, Cambridge, MA.

Gopnik, A. (2009). The philosophical baby. — Houghton Mifflin, New York, NY.

Graham, J., Haidt, J. & Nosek, B. (2009). Liberals and conservatives rely on different sets of moral foundations. — J. Personal. Soc. Psychol. 96: 1029-1046.

Greene, J.D. (2007). The secret joke of Kant's soul. — In: Moral psychology, Vol. 3: the neuroscience of morality: emotion, disease, and development (Sinnott-Armstrong, W., ed.). MIT Press, Cambridge, MA, p. 35-80.

Haidt, J. (2012). The righteous mind: why good people are divided by politics and religion. — Pantheon, New York, NY.

Haidt, J. & Joseph, C. (2004). Intuitive ethics: how innately prepared intuitions generate culturally variable virtues. — Daedalus 133: 55-66.

Haidt, J. & Joseph, C. (2007). The moral mind: how 5 sets of innate moral intuitions guide the development of many culture-specific virtues, and perhaps even modules. — In: The innate mind, Vol. 3 (Carruthers, P., Laurence, S. & Stich, S., eds). Oxford University Press, Oxford, p. 367-391.

Haidt, J., Koller, S.H. & Dias, M.G. (1993). Affect, culture, and morality, or is it wrong to eat your dog? — J. Pers. Soc. Psychol. 65: 613-628.

Harman, G. (1999). Moral philosophy meets social psychology: virtue ethics and the fundamental attribution error. — Proc. Aristotel. Soc. 99: 315-331.

Hsu, M., Anen, C. & Quartz, S.R. (2008). The right and the good: distributive justice and neural encoding of equity and efficiency. — Science 320: 1092-1095.

Hume, D., Norton, D.F. & Norton, M.J. (1739/2007). A treatise of human nature. — Oxford University Press, Oxford.

Jensen, K., Call, J. & Tomasello, M. (2007). Chimpanzees are rational maximizers in an ultimatum game. — Science 318: 107-109.

Jensen, K., Hare, B., Call, J. & Tomasello, M. (2006). What's in it for me? Self-regard precludes altruism and spite in chimpanzees. — Proc. Roy. Soc. Lond. B: Biol. Sci. 273: 1013-1021.

Kelly, D. (2011). Yuck!: the nature and moral significance of disgust. — MIT Press, Cambridge, MA.

Kierkegaard, S., Evans, C.S. & Walsh, S. (1937 [2006]). Fear and trembling. — Cambridge University Press, Cambridge.

Kitcher, P. (2011). The ethical project. — Harvard University Press, Cambridge, MA.

Kitcher, P. (2014). Is a naturalized ethics possible? — Behaviour 151: 245-260.

Knapp, C. (2003). De-moralizing disgustingness. — Philos. Phenomenol. Res. 66: 253-278.

Kohlberg, L. (1973). The claim to moral adequacy of a highest stage of moral judgment. — J. Philos. 70: 630-646.

Koleva, S.P., Graham, J., Iyer, R., Ditto, P.H. & Haidt, J. (2012). Tracing the threads: how five moral concerns (especially Purity) help explain culture war attitudes. — J. Res. Personal. 46: 184-194.

MacIntyre, A. (1981 [2007]). After virtue, 3rd edn. — University of Notre Dame Press, Notre Dame, IN.

Nussbaum, M. (2004). Hiding from humanity: disgust, shame and the law. — Princeton University Press, Princeton, NJ.

Obeyeseker, G. (2002). Imagining karma: ethical transformation in Amerindian, Buddhist, and Greek Rebirth. — University of California Press, Berkeley, CA.

Peterson, C. & Seligman, M. (2004). Character strengths and virtues: a handbook and classification. — Oxford University Press, Oxford.

Plakias, A. (2013). The good and the gross. — Ethic. Theor. Moral Pract. 16: 261-278.

Proctor, D., Williamson, R.A., de Waal, F.B. & Brosnan, S.F. (2013a). Chimpanzees play the ultimatum game. — Proc. Natl. Acad. Sci. USA 110: 2070-2075.

Proctor, D., Williamson, R.A., de Waal, F.B. & Brosnan, S.F. (2013b). Reply to Jensen et al.: equitable offers are not rationally maximizing. — Proc. Natl. Acad. Sci. USA 110: E1838.

Quine, W.V. (1969). Epistemology naturalized. — In: Ontological relativity and other essays. Columbia University Press, New York, NY

Quine, W.V. (1979). On the nature of moral values. — Critic. Inq. 5: 471-480.

Rawls, J. (1951). Outline for a decision procedure for ethics. — Philos. Rev. 60: 177-197.

Rawls, J. (1971). A theory of justice. — Harvard University Press, Cambridge MA.

Rheingold, H. (1982). Little children's participation in the work of adults, a nascent prosocial behavior. — Child Dev. 53: 114-125.

Sellars, W., Rorty, R. & Brandom, R. (1997). Empiricism and the philosophy of mind. — Harvard University Press, Cambridge, MA.

Shweder, R.A. (2002). The nature of morality: the category of bad acts. — Med. Ethics 9: 6-7.

Shweder, R.A., Much, N.C., Mahapatra, M. & Park, L. (1997). The "big three" of morality (autonomy, community, and divinity), and the "big three" explanations of suffering. — In: Morality and health (Brandt, A. & Rozin, P., eds). Routledge, New York, NY, p. 119-169.

Singer, P. (1972). Famine, affluence, and morality. — Philos. Publ. Affairs 1: 229-243.

Thagard, P. & Verbeurgt, K. (1998). Coherence as constraint satisfaction. — Cogn. Sci. 22: 1-24.

Thompson, M. (1995). The representation of life. — In: Virtues and reasons: Philippa Foot and moral theory (Hursthouse, R., Lawrence, L. & Quinn, W., eds). Oxford University Press, New York, NY, p. 247-296.

Warneken, F. & Tomasello, M. (2006). Altruistic helping in human infants and young chimpanzees. — Science 311: 1301-1303.

Wheatley, T. & Haidt, J. (2005). Hypnotic disgust makes moral judgments more severe. — Psychol. Sci. 16: 780-784.

Williams, B. (1972). Morality: an introduction to ethics. — Cambridge University Press, New York, NY.

Wilson, E. (1975 [1980]). Sociobiology: the new synthesis. — Harvard University Press, Cambridge, MA.

Wolf, S. (1982). Moral saints. — J. Philos. 79: 419-439.

Wong, D. (2006). Natural moralities: a defense of pluralistic relativism. — Oxford University Press, Oxford.

[When citing this chapter, refer to Behaviour 151 (2014) 229–244]

Human nature and science: A cautionary essay

Simon Blackburn [a,b,*]

[a] Faculty of Philosophy, Trinity College, University of Cambridge, Sidgwick Avenue,
Cambridge CB3 9DA, UK
[b] Department of Philosophy, University of North Carolina, 108B Caldwell Hall,
240 East Cameron, Chapel Hill, NC 27599-3125, USA
[*] Author's e-mail address: swb24@cam.ac.uk

Accepted 15 October 2013; published online 5 December 2013

Abstract
In this paper I draw upon the philosophical tradition in order to question whether scientific advances can show us as much about human nature as some may expect, and to further question whether we should welcome the idea that scientific interventions might improve that nature.

Keywords
convention, culture, design, education, emotion, epigenetics, ideology, morality, philosophy, science, transhumanity.

1. Introduction

Questions of human nature are as old as philosophy itself. Perennials include whether there is such a thing as free-will, or whether there is such a thing as disinterested behaviour, or altruism, or public spirit. We ask about the relative importance of reason and emotion. We ask if the individual precedes society, or is just an abstraction from it. Many disciplines join in. Social psychologists ask whether men are more aggressive than women, whether violence on television affects those who watch it, or why human beings enjoy watching it in the first place. Evolutionary psychologists speculate about our hominid ancestors in their pleistocene environments, or pick up clues about what we are like by comparing us to chimpanzees and bonobos. Geneticists explore the heritability of various traits, while neuroendocrinologists study the extensive influence of our hormonal systems on our behaviour. It is tempting to suppose that ongoing science will completely eclipse whatever speculations or arguments the philosophical tradition once brought to these issues. The laboratory overturns the armchair, and although philosophers have fought back (Kitcher, 1985) triumphalism about this is prevalent to the point of orthodoxy.

Kant distinguished between physiological knowledge of a human being, which is what nature makes of the human being, and pragmatic knowledge, which is the investigation "of what he as a free-acting being makes of himself, or can and should make of himself". Books about biology and physiology are about the first, but books about human nature are about the second. It is our psychological traits that are in question, and it is the importance of those traits that give the question its apparent urgency. For we also suppose that our theories about ourselves matter. Clearly, if I believe that everyone is ultimately selfish, I will conduct my life differently, and may myself become selfish, untrusting and untrustworthy, and other people may follow suit. Around the beginning of the eighteenth century philosophers such as Shaftesbury (1977) lamented the damage done by the pessimistic view of human depravity characteristic of Calvinism and other versions of Christianity. The issue of egoism versus altruism became centre-stage, with optimists like Shaftesbury himself, Butler (1953) and Hutcheson (1725) pitted against pessimists such as Hobbes (1996) and Mandeville (1988). Adam Smith's two books seem to sit on the fence, or rather, one is apparently on one side, and the other on the other (Otteson, 2000). One thing both sides agreed on was that there is a real, urgent question here (Gill, 2006).

2. Human nature and human artifice

Nevertheless, we should ask at the outset whether the concept of a human nature is itself merely a remnant of the Aristotelian, essentialist, idea that everything has a Natural State. The problem with this essentialism is that there is variation in the human genome — indeed, it is generally supposed that fomenting this variation is the function of sexual reproduction and its accompanying genetic recombination — and furthermore the journey from genome to phenotype shows no one natural relationship. It only shows a variety of 'norms of reaction', as genes express themselves differently in different environments. There is no reason to doubt that a highly mixed population, in respect of altruism, greed, public spirit, aggression and any other interesting trait, could have evolved and could be stable, just as a highly

mixed population in terms of height or weight has done. All the theorist of
human nature can hope for is that there are interesting constancies, just as
there are other phylogenetic constancies in animal development. But there
will be no a priori certainty that even apparent constancies are not in principle
environmentally plastic. Every stage of the journeys from genes to proteins
to brains to psychologies may be highly variable. Even twin studies would
be of little help here, since from conception twins share the same bodily
environment for nine months, and by global and historical standards they
share pretty identical cultural environments thereafter.

One idea might be that we try to imagine ourselves stripped of the overlay
of civilization and culture, to get down to the raw original hominid nature.
But whatever we may think about human nature in the raw, and whatever our
optimism that there is such a thing as the raw, one of the inescapable facts
of human existence in a natural world is that we have evolve concepts and
strategies for dealing with it. We did not make the natural world but just as
birds construct nests, so we construct our own environmental niches, and in
particular the social niches in which we cohabit with conspecifics.

The first philosopher to explore this mechanism thoroughly was David
Hume (1739), who argued convincingly that many of the most important
facets of our social lives, such as the conventions underlying language, trust,
law, contracts, property, money, and government, involve institutions and
structures that are in some sense 'artificial' or socially constructed. Hume is
not overly concerned about whether they are the result of intentional design,
or might have evolved simply because of their functional utility, but in either
event they depend upon a conditional willingness of each agent to play his or
her part provided the others do so. Hume also thought that although it is only
a selfish concern, together with a reasonable farsightedness, that motivate us
to put these conventions in place, by establishing them we build a 'vault'
or protection that shelters us from the worst effects of that selfishness. They
therefore provide a very effective substitute for any more general altruism
or benevolence. This enabled him to dismiss the question that separated
optimists from pessimists, of the 'dignity or meanness' of human nature.
It is not so important whether human beings are egoists or altruists if either
way we can end up with social and political solutions to which we can all
adhere (Gill, 2006). Hume himself was a cheerful moderate about human
nature, supposing that we need only that "there is some particle of the dove
mingled into our frame along with the elements of the wolf and the serpent"

in order both to explain and to justify our moral sentiments (Hume, 1998: 147). It is our capacity for finding institutional solutions to the problems of living together that substitutes for any imagined golden age, and wards off the dangers inherent in our limited capacities for altruism.

Hume's constructive story is a marked improvement over that of Hobbes, who thought that the only remedy for the 'war of all against all' which made up the state of nature, was that people should surrender all use of force to an absolute sovereign. This solution prompted John Locke's famous rebuttal that "This is to think that men are so foolish that they take care to avoid what mischiefs may be done them by polecats or foxes, but are content, nay, think it safety, to be devoured by lions". Hume needs no sovereign, and he carefully avoids any mention of contract in his story: as Hobbes himself saw, contracts mean nothing in his state of nature, and respect for contracts is one of the end products of the conventions he describes, not a presupposition of them.

There will sometimes be immediate temptations to defect from such conventions. So it is notable that a large part of moral education involves drumming their inviolable nature into children. We might plead with children to be nicer to each other, with more or less success. But we pretty much order them to respect property, not to lie, to keep promises, or to obey the law. In these areas defection brings retribution. It is of course contingent whether we succeed, and science can certainly explore the contours of any emotional or cognitive deficits that prevent the education from being successful with any particular subject. But Hume stands in the way of supposing that any failure should be interpreted as the fault in some specific 'moral' faculty or innate moral module (Hauser, 2006; for a longer rebuttal see Blackburn, 2008). Insofar as the conventions are artificial or constructed, so is respect for them. It will not be an innate gift so much as the conventional implementation of a more general trait, such as sensitivity to the applause or criticism of other people. There may, of course, be other innate emotional traits that assist this, such as sensitivity to fair treatment, and a propensity to resentment when it is denied (de Waal, 2006). These were particularly emphasised by Hume's successor Adam Smith (1982). Eventually the imagined voice of others becomes our own internal sanction, Smith's man within the breast representing what an ideal spectator would say or feel about our actions. But allowing this mechanism is not to deny that we may be assisted to behave better when there are others actually watching us, or even supposed or just vividly imagined to be watching us (Bateson et al., 2006).

3. Real science and scientific ideology

Hume's pioneering exploration answers the brash belief that there is something 'unscientific' about invoking culture or environment as potential determinants of a resulting human nature — albeit a nature that may well vary from culture to culture. Appealing to culture is not appealing to some kind of cloudy super-organism, something like a Hegelian Spirit of the Age, conceived of as a weird and ghostly causal force above and distinct from the world around us. It is just appealing to the very important and pervasive part of the environment that is due to the doings of other people. But Hume was a historian, and his laboratory was his library. If years in the library or with anthropologists in the field do not throw up stable answers to the old perennial questions, answers to which all unbiased investigators might hope to converge, can experimental methods do better? Perhaps, and they can certainly assist us. But I first want to sound one cautionary note. I think we must always be careful of confusing science with what we might call the ideology of science. For sad to say even distinguished scientists are capable of writings that are not themselves distinguished science. They do so when they read ideologies into the natural record, rather than read observations out of it, or, more insidiously, when metaphors and analogies are allowed to get out of hand. The unfortunate notion of the selfish gene is a paradigm case (Dawkins, 1976). Although the suggestion of a particular motivation guiding genes and their evolution was never intended as more than a metaphor, there is no doubt that it has taken on a life of its own (nor any doubt that it had a symbiotic relationship with the increasing monetisation of the entire field of human relationships, that accelerated so horrifically at just the time of its publication).

The problem of the evolution of altruism, defined in the biologist's sense in terms of a transfer of reproductive fitness from one organism to another, is often posed as one of individual selection versus group selection. But this is wrong. Altruism can evolve wherever the benefits donated by the altruists tend to fall on others who are likely to breed altruists. The sorting of a population into 'groups' is one way this can happen, but correlated interactions can occur in other ways as well, just so long as "the benefits of the acts fall disproportionately on those who are likely to pass the behaviour on" (Godfrey-Smith, 2010: 120). This generates a quite general formal framework, of which two of the most familiar mechanisms, kin selection and reciprocity, are but two examples. In 1975 William Hamilton himself recognised that kinship should be considered just one way of getting positive regression of genotype in the recipient of [altruistic behaviours] and it is this positive regression that is vitally necessary for altruism (quoted in Godfrey-Smith, 2010: 176).

Obviously enough, from the fact that an organism is of a type that has had to survive and to evolve, we cannot at all deduce that it has to care about nothing but its own survival, or its own 'interest' or the number of its own progeny or their fecundity, or any other single thing. As many writers have pointed out the inference from function to overt psychology is simply not available. Sliding over the gap is exactly like inferring from the fact that our sexual drives have an evolutionary function, namely procreation, that all we want when we want sex is to procreate. This is an inference that generations of churchmen have wished were valid, but happily for human pleasures, and for the pharmacology industry, it is not (Blackburn, 2004). William James is a useful ally here: "Not one man in a billion, when taking his dinner, ever thinks of utility. He eats because the food tastes good and makes him want more" (James, 1902: 386).

Hence, we might in principle be altruistic in just the same spirit as the man eating his dinner. For there are plenty of evolutionary dynamics in which individuals who bear costs in order to assisting their kin, their neighbours, those who help them, or the collective, do better than those who do not. Nice guys sometimes do finish first, and this should be no more surprising than that less lethal parasites flourish rather than their greedier, but lethal, cousins — the dynamic whereby diseases such as maxymatosis in rabbits tend to become less lethal over time. By the same mechanism, as the Horatii taking their oath no doubt realised, in a world in which we must all hang together or else we all hang separately, those who are adapted to hang together do best.

4. Education, emotion and morality

The misreading of Darwin is perhaps the most prevalent, and dangerous, misinterpretation of science that philosophers can help to fight, but I now turn to making some remarks about neurophysiology and our understandings of 'what human beings make of themselves or can and should make of themselves'. I shall illustrate the point I want to make by remarking on a famous result, one that has played an important role in bringing epigenetics out of the dustbin into which it was placed by genetic triumphalism ten to fifteen years ago, and back into the central position it now occupies.

It was indeed neurophysiological work that firmly brought culture back into the picture. At McGill University researchers subjected rats to different regimes of maternal behaviour (Meaney, 2001; Weaver et al., 2004). It was found that: "the mother's licking activity had the effect of removing dimmer switches on a gene that shapes stress receptors in the pup's growing brain. The well-licked rats had better-developed hippocampi and released less of the stress hormone cortisol, making them calmer when startled. In contrast, the neglected pups released much more cortisol, had less-developed hippocampi, and reacted nervously when startled or in new surroundings. Through a simple maternal behaviour, these mother rats were literally shaping the brains of their offspring. In short licking and grooming release serotonin in the pup's brain, which activates serotonin receptors in the hippocampus. These receptors send proteins called transcription factors to turn on the gene that inhibits stress responses". And as with rats so, we can be sure, with us. If maternal behaviour and diet can make these changes, then so much for any brash confidence that our genes are our fate, and that culture is causally inert.

Epigenetics suggests that our genetic inheritance is only the beginning, since transcription factors influencing the switching on or off of these mechanisms of interaction are subject to environmental influence, and this at different levels, from inside the cell, and possibly at levels below that, up to what I have already called culture. We now know that even diet can influence the methylation of genes, and this in turn affects their expression and suppression.

The point I want to make in this essay is that it should not have actually needed the neurophysiology. The connection between maternal licking and less stressed offspring is entirely visible at the macroscopic, observational level. It could have been established purely empirically. Of course then, there would have been a placeholder: it would have been supposed that there was 'some difference' in the brain of the neglected rats that caused them to be more nervous and stressed than those who were properly mothered. It would not have been known that hippocampi, cortisol, and transcription mechanisms were involved, and it is certainly important to know those things. But you can clearly know the upshot without knowing how it is implemented or realized in the neural structures that subsume it. A Kant wanting to know what rats can and should make of themselves could have observed and made use of the result. Yet in the heyday of genetic determinism that followed the unravelling of the human genome, the phenotypical observations probably needed the data from the neural underpinnings to themselves be noticed (or believed).

Moral theory has been affected by more recent misinterpretations of the relation between science and ethics. This comes from work with fMRI scanners. These show, for instance, that more emotional changes, and more emotional centres in the brain are involved when people contemplate some 'forbidden' scenarios — such as pushing the well-known fat man off a bridge in front of a train — than when we are simply calculating the utility of outcomes where no deep prohibitions are in play. I am quite happy to believe that this is so, and indeed would be surprised were it not: it is, after all, quite scary to imagine with any detail what it would be like to push a fat man under a train. But it has then been inferred that this is a reason for preferring outcome-based, consequential or utilitarian reasoning to the emotionally contaminated reasoning involved with notions of duty and obligation, prohibition and permission (Greene, 2003, 2008; Singer, 2005). Yet there are at least two things wrong with this inference. First, it generalises wildly from just one example of the contrast. Yet it is easy to think of others where emotion may attach more to thinking of the consequences than the duties involved. For instance, I might think to myself that I should invite John to my party, because he invited me to his (duty). But this does not make me break into a sweat, or make my heart pound, and I would be very surprised if it makes the emotional centres of my brain go into overdrive. The judgment of duty is made in an entirely clinical frame of mind. On the other hand when I think of the consequences of John coming to the party — his likely drunken behaviour for instance — I might very well start to sweat and worry. So any association between thinking in terms of duty and feeling emotional, or between thinking of consequences and feeling unemotional, is entirely coincidental. But secondly, and more importantly, even if the association we more firm than I am suggesting, it would entirely fail to show that either kind of thinking is in any sense second-rate or dispensable, and supposing that it does so is simply a residue of an ideology of mistrust of the emotions, and the corresponding worship of 'reason' as an independent vector or force in our decision making (Berker, 2009). Hume of course was in the forefront of opposing that ideology as well, and he has been thoroughly vindicated insofar as some philosophers and other neurophysiologists believe emotions to be the important drivers of all our practical concerns (Damasio, 1995). And I would add as a rider that the presence or absence of an emotional phenomenology, which is doubtless the mental correlate of the kinds of excitation that neurophysiologists can detect, does nothing to settle any pressing questions about the nature of moral judgment, as Hume also saw when he insisted on the category of the 'calm passions', or motivational dispositions that have no particular phenomenology or 'feel'.

The general message is that it takes a whole picture, a picture of the entire agent behaving in different contexts over periods of time, to cement an interpretation in place. And this may make us pause over how much understanding of ourselves is likely to come out of the neurophysiology laboratory. It suggests a caution about what can be expected from neurophysiology, brain scans, and the rest. We may indeed discover endless fascinating things about the mechanisms involved in us doing what we do. But we will not, thereby, discover ready new interpretations of what it is that we do nor new ways of evaluating them. The neurophysiology itself needs interpretation, and the only way of providing that is by calibrating it against the common behaviour that is the starting point of interpretation in any event. Brain writing, even if it were salient and stable and reidentifiable and common to different people, would need interpretation and evaluation as much as any other text.

To illustrate this imagine a replicated, definite result along the lines that when women read a map one bit of the brain is active, whereas when men read a map a different bit is. Would that show that women 'read maps differently' from men, or are better or worse at it, or do not understand maps in the same way? Of course not: those interpretations would need to earn their keep at the tribunal of capacities and behaviour. If the capacities and behaviour are identical, then what we have learned is that different bits of the brain can subsume them, not that the different bits of the brain are subsuming different abilities or understandings. And if the capacities and behaviour are different — if women get lost more easily than men for instance — well then, we learn that out in the field, and in fact without the resort to fMRI scans or other high-tech knowledge.

For a different actual example, consider the well-known and fascinating results of Antonio Damasio and his collaborators on the deficits exhibited by prefrontal patients, that is, people with damage to the ventromedial prefrontal cortex. The clinical observation is that such patients, while they may be perfectly normal in respect of perception, memory, language skills, and logic, may also be very bad at practical decision-making, and this in turn makes them significantly worse at living any kind of normal life. Damasio's own interpretation of their deficit was in terms of an emotional disability, construed as the inability to 'somatically mark' actual and projected events, and thereby attach a positive or negative valency to them. And this in turn seems to lend some support both to William James's old theory of emotions as a kind of perception of bodily arousal, and to give help to those supposing that emotion has a greater role in practical decision making than its nebulous competitor, reason.

However there are rival interpretations of the data: one is that prefrontal patients lack the skill to perform a kind of 'time travel', both remembering past experiences with themselves as the subject, in the way normal people might remember eating a nice or nasty meal, and imagining future experiences with themselves as the subject, in the way normal people can easily imagine eating the same meal again, and then anticipating the episode with either pleasure or aversion (Gerrans, 2007). I offer no opinion on the debate, but merely cite it as an example where the facts of behaviour, and facts about their neural substrate, may be relatively clear, but the question of interpretation very open indeed.

A final well-known example, which I shall only mention very briefly, derives from work by Benjamin Libet and others on the interesting facts concerning the way excitation in the motor cortex (the so-called readiness potential) precedes, by a third of a second or more, the time at which an agent believes himself to have decided to act. Other results extend the time interval; yet others suggest that the readiness potential does not after all indicate that a decision has been made (Trevena & Miller, 2009). The implication that a little philosophy could have warded off was that such results undermined any notion of free-will. It could only do so if this were thought of as necessarily connected with an intervention into the physical world from the distinct domain of mind or consciousness. But in contemporary philosophy this Cartesian conception of free-will, while not perhaps entirely dead, is more an object of ridicule than of serious discussion. It is neither true to the phenomenology of normal action, nor has it anything to contribute to the ordinary business of allocating responsibility for actions, or accepting excuses for them.

These cautionary remarks about the interpretation of neurophysiological results need not extend to the psychology lab. The psychologist dealing with capacities and behaviour can indeed tell us whether women read maps as quickly and efficiently as men, or under what circumstances a group playing iterated prisoners' dilemma games falls into a cooperative or a noncooperative equilibrium. It is of course an urgent and difficult question to settle the levels at which psychological categories and neurobiological explanations interact, and so to decide whether different impairments or associated impairments are best explained in terms of a psychological inability (for instance, the lack of a theory of mind, or the lack of some imaginative capacity) or better interpreted not in terms of a psychological architecture at all, but purely in terms of neuronal or other abnormalities. Here, the utility of the resulting theories rather than any antecedent plausibility will turn out to be decisive.

5. A trans-human future?

I will conclude by turning to the idea of deploying genetic manipulations to bring about a psychologically improved human being: the old ambition of eugenics in its new scientific dress. Can we set about beautiful design, where nature has merely kludged? Again, resources from the philosophical tradition have something to say.

If the question is whether we can change human nature, then the blunt response has to be that we already know how to do so, one person at a time. We do it rather like rats do it, by licking and grooming. We do it when we socialize our children, when we teach them our language, when we introduce them to property and promises, forbearances and co-operations, conventions and norms, and the millions of little capacities that eventually fit them for their lives as grown-ups. This is an agonizing process, as parents know, and we endlessly debate and tinker with the choices in front of educators. We do not even know the best way to teach children to read, for example, and of course there may be no such thing as the best way: only a multitude of ways, some more suitable for some, and others for others.

That is cultural influence. But pharmacological interventions as well can change our nature: it is widely supposed that large amounts of hallucinogenic drugs can cause or trigger schizophrenia, for example. Large doses of Ritalin make naughty children docile, at least for a time. And presumably, as eugenicists hoped, selective breeding, or genetic engineering, might in principle eventually change the gene pool, and lead to different kinds of people.

Should we be frightened at these prospects? Again I am not talking about the elimination or suppression of physical disease and illness to which some simple genetic peculiarity makes some people prone: Huntington's chorea and Duchenne's muscular dystrophy being perhaps the best publicly known examples. Few people could quarrel with interventions that eliminate those. Some people even talk of the ageing process as a kind of disease that a suitable genetic modification might halt or even reverse, giving us a fountain of perpetual youth. This is more radical, and there would doubtless be much discussion the pros and cons of such an intervention, were it on the horizon.

But it is psychologies that interest us when we talk of human nature. We are talking of changes that might give us large-scale, heritable psychological dispositions that improve us beyond our ancestors. So the question is whether we can imagine a genetic engineering not to remove deficits and disease, but to improve human nature, rather as eugenic programs were hoped to do so around a century ago. We could imagine enhancements, producing people who are, for instance, more just, less selfish, more courageous, more intelligent, more imaginative, more prudent, more humorous and better company. Of course the aims of old eugenic programs might strike us as comical when they were not wicked, but I am sure that today many people congratulate themselves on being able to do better. They, in the old days, ushered in a nightmare, but we, in the twenty-first century, will wave a wand. We know how to aim at utopia, and we are responsible enough to do so.

This is a brave, can-do attitude, but, I shall argue, not at all one we should share (Kitcher, 1996). First of all, there are obvious reasons for scientific caution. There are very few aspects of normal human development, and especially brain development, that are not polygenic. So we immediately face a combinatorial explosion: if each of our 25 000 or so genes can interact in a great number of ways with even a small proportion of the remaining 24 999, the underlying processes will number in the countless millions, and the prospect of understanding them in fine detail is proportionately remote. The project of unravelling the genome itself, so widely applauded, and heralded as ushering in the brave new dawn, was obviously trivial by comparison. And that is not even to begin on the kinds of epigenetic factors already described, whose variation introduces a whole new dimension of complexity and uncertainty.

But once more it is the resources coming from philosophy that interest me. There are two from ancient philosophy that are particularly relevant. One is the Socratic doctrine of the unity of the virtues: the idea that one cannot simply be courageous, or just, or generous or merciful tout court. Exercising one virtue requires exercising others. If courage is not to be mere stupidity or mere lack of imagination for example, it needs to be accompanied by awareness and judgement; equally if it is not to go over the top into mad rashness, then it has to be combined with caution and prudence. Similarly for other virtues. A judge or minister of state cannot be merciful without being courageous, since sometimes that will be required in the face of the anger of people clamouring for punishment and death. And so on across the board.

This alone makes any project of 'improving' human nature with genetic intervention highly fraught. Do you want kinder, more generous people? We try to get them by bringing people up in kind environments, gently rewarding exhibitions of kindness, discouraging the reverse. But might we have a magic genetic bullet, bypassing this troublesome cultural exercise? Well, misjudged kindness is a bad thing: kind parents often spoil their children. They may smother them, patronize them, take away their initiative, infantilize them, or prevent their maturity. As the Socratic doctrine insists, proper kindness can only coexist with judgment, tact, imagination, respect for the dignity of others, and a host of other nuanced virtues. And even then there will be little consensus on when just the right mix has been achieved. All we can do is aim in general directions, and then keep our fingers crossed.

But when we look at more detail, in the world as we have it, the picture becomes yet more complicated, in a way that also involves the second classical resource: the Aristotelian doctrine of the mean. Consider a virtue such as generosity. Most people are generous some of the time. Suppose we found a genetic intervention that magnified the trait. This might sound to be a desirable breakthrough, until we ask how much we want. Promiscuous generosity makes a person a soft touch, a pushover, and since he will be a poor trustee of his own interest, he is bound to be a nuisance to many of those around him. So we need something between being miserly and niggardly on the one hand, and imprudent and reckless on the other hand. And once more the Socratic doctrine applies: justice is just one of the traits that have us being rather less generously disposed to some people than others, as anyone who has to mark examination efforts well knows. Of course what we would like is someone who is kind and generous to the right degree in the right circumstances. We want a relatively plastic disposition, and one that goes along with consider-able context-sensitivity, but imagining this difficult bulls-eye hit by a magic bullet is a fantasy. It is imagining someone navigating the social world with-out experience, which is no more likely than navigating an unfamiliar city or coastline also without experience.

Even if we confine ourselves to that old and apparently uncontroversial aim of eugenics, 'improved intelligence' parallel worries arise. Intelligence can famously be used for good or bad, strategically or cooperatively. Wily Odysseus was the most intelligent of Agamemnon's men, much admired for his deceptions and frauds, lies and plots. And it too is subject to a host of complexities. Even within the charmed meadows of academic life, we all know colleagues, brilliant people in their field, whom we would not trust very far outside that field. Until researchers find a gene for getting hold of the wrong end of the stick, and manage to remove it from the population without any other side-effects, I think we should not hold our breath.

All these remarks are about the nature of the aim, and entirely abstract. But when we turn to imagining the possible implementation of genetic in-terventions, things get much worse. We would be naïve if we thought that new genetic designs, 'synthetic biology', would be immune to catastrophic mistakes or catastrophic misuse: error or terror as it has been called. We would also be naïve if we thought a genetic utopianism would be immune to the normal pressures of politics and commerce. The forces in charge of propelling us towards the future are not future forces, but present ones. We know that the eugenicists of the twentieth century were, for instance, Aryan racial supremacists, and had a very peculiar view of human excel-lence. Who would like to guess what those of this and subsequent centuries might be? Capitalism would like an intervention promoting consumer envy and lifestyle discontent. People on the Right would like one diminishing con-cern for social justice, while those on the Left would like one for increasing it. Pharmaceutical companies would not pay for research to make childhood naughtiness disappear without their continued help, and while moralists put in their bid for kindness and intelligence, the Pentagon will press for a less compassionate and more docile soldiery. In other words, we must beware of the idea that there is an abstraction called 'Science', an insightful, imagina-tive, impersonal, just, benevolent, invisible hand and that we can leave the people of the future to it. There is no such hand: as William James also said, the trail of the human serpent is everywhere. And of course, given that in the foreseeable future genetic interventions look like being very expensive, we might not entirely welcome a world in which the wealthy get something approaching a disease-free immortality, with wonderful brains thrown in, and the rest go hang.

I have spent much of my time sounding cautionary notes. But let me finish on a more upbeat theme. I find it immensely interesting and encouraging that the old Delphic injunction: 'know thyself', now rules so much scientific activity, as well as its old domain, the humanities. There are no trade-union barriers in the pursuit of knowledge. The investigation of what we can and should make of ourselves is, as I began by saying, as old as philosophy. It is as old as Homer and Augustine, Milton, Hume or James, and as new as the latest ideas from game theorists, evolutionary psychologists, neurophysiologists, pharmacologists, zoologists, economists, and perhaps even quantum theorists and engineers. It is hard work listening to everyone, but all I hope to have done is to suggest that there is still a place for philosophers and their traditions at the table.

References

Bateson, M., Nettle, D. & Roberts, G. (2006). Cues of being watched enhance cooperation in a real-world setting. — Biol. Lett. 2: 412-414.

Berker, S. (2009). The normative insignificance of neuroscience. — Philos. Publ. Affairs 37: 293-329.

Blackburn, S. (2004). Lust. — Oxford University Press, New York, NY.

Blackburn, S. (2008). Response to Marc Hauser's Princeton Tanner Lecture, available online at http://www2.phil.cam.ac.uk/~swb24/PAPERS/Hauser.pdf

Butler, J. (1953). Fifteen sermons preached at the Rolls Chapel. — Bell & Sons, London.

Damasio, A. (1995). Decartes's error. — Harper Collins, New York, NY.

Dawkins, R. (1976). The selfish gene. — Oxford University Press, Oxford.

de Waal, F.B.M. (2006). Primates and philosophers. — Princeton University Press, Princeton, NJ.

Gerrans, P. (2007). Mental time travel, somatic markers and 'Myopia for the future'. — Synthese 159: 459-474.

Gill, M. (2006). The British moralists on human nature and the birth of secular ethics. — Cambridge University Press, Cambridge.

Godfrey-Smith, P. (2009). Darwinian populations and natural selection. — Oxford University Press, Oxford.

Greene, J.D. (2003). From neural 'is' to moral 'ought': what are the moral implications of neuroscientific moral psychology? — Nature Rev. Neurosci. 4: 847-850.

Greene, J.D. (2008). The secret joke of Kant's soul. — In: Moral psychology 3: the neuro-science of morality: emotion, brain disorders, and development (Sinnott-Armstrong, W., ed.). MIT Press, Cambridge, MA, p. 35-79.

Hauser, M. (2006). Moral minds: how nature designed our universal sense of right and wrong. — Ecco Press, New York, NY.

Hobbes, T. (1996). Leviathan (Tuck, R., ed.). — Cambridge University Press, Cambridge.

Hume, D. (1888 [1739]). A treatise of human nature (Selby-Bigge, L.A., ed.). — Oxford University Press, Oxford.

Hume, D. (1998 [1751]). An enquiry concerning the principles of morals (Beauchamp, T.L., ed.). — Oxford University Press, Oxford.

Hutcheson, F. (1725). An inquiry into the original of our ideas of beauty and virtue. — J. Darby, London.

James, W. (1902). The principles of psychology v. II. — Henry Holt, New York, NY.

Kitcher, P. (1985). Vaulting ambition: sociobiology and the quest for human nature. — MIT Press, Boston, MA.

Kitcher, P. (1996). The lives to come. — Touchstone, New York, NY.

Mandeville, B. (1988 [1924]). The fable of the bees (Kaye, F.B., ed.). — The Liberty Press, Indianapolis, IN.

Meaney, M. (2001). Maternal care, gene expression, and the transmission of individual differences in stress reactivity across generations. — Annu. Rev. Neurosci. 24: 1161-1192.

Otteson, J. (2000). The recurring Adam Smith problem. — Hist. Phil. Q. 17: 51-74.

Shaftesbury, 3rd Earl of (1977 [1699]). An inquiry concerning virtue or merit (Walford, D., ed.). — Manchester University Press, Manchester.

Singer, P. (2005). Ethics and intuitions. — J. Ethics 9: 331-352.

Smith, A. (1982 [1759]). The theory of moral sentiments (Raphael, D.D. & Macfie, A.L., eds). — The Liberty Press, Indianapolis, IN.

Smith, A. (2000 [1776]). An inquiry into the nature and causes of the wealth on nations (Canaan, E., ed.). — The Liberty Press, Indianapolis, IN.

Trevena, J. & Miller, J. (2010). Brain preparation before a voluntary action: evidence against unconscious movement initiation. — Consci. Cogn. 19: 447-456.

Weaver, I.C.G., Cervoni, N., Champagne, F.A., D'Alessio, A.C., Sharma, S., Seckl, J.R., Dymov, S., Szyf, M. & Meaney, M.J. (2004). Epigenetic programming via maternal behaviour. — Nature Neurosci. 7: 847-854.

[When citing this chapter, refer to Behaviour 151 (2014) 245–260]

Is a naturalized ethics possible?

Philip Kitcher *

Department of Philosophy, Columbia University, 708 Philosophy Hall, MC 4971,
1150 Amsterdam Avenue, New York, NY 10027, USA
*Author's e-mail address: psk16@columbia.edu

Accepted 15 September 2013; published online 27 November 2013

Abstract
I offer an account of the evolution of ethical life, using it to elaborate a meta-ethical perspective and a normative stance. In light of these discussions, I attempt to answer my title question.

Keywords
ethics, evolution, ethical progress, ethical method, naturalistic fallacy.

1. Introduction

Philosophical thought about ethics is typically dominated by the vision of a 'theory of everything ethical', some grand system of principles that would yield an unambiguous decision about the goodness of every state of affairs or the rightness of every action (see, for a prominent recent example, Parfit, 2011). Within that perspective, ethical naturalism consists in connecting central ethical predicates — 'good' or 'right', say — to natural properties. So utilitarians declare that states are good insofar as they maximize the aggregate balance of pleasure over pain across all sentient creatures (Bentham, 1988), and ambitious sociobiologists propose that actions are right if they promote the proliferation of human DNA (Ruse & Wilson, 1986). Views of this sort are routinely criticized on the grounds that they not only fail to deliver the correct ethical judgments about particular cases, but also are guilty of the primal sin of inferring judgments of value from judgments of fact (the so-called 'naturalistic fallacy').

Whether naturalism, as so conceived, can answer its critics is not my concern. For, in my view, the entire framework in which ethical discussions are typically cast is faulty. There are excellent reasons for supposing that the idea of a *scientific* 'theory of everything' is a delusion (Dupre, 1993; Cartwright, 1999; Kitcher, 2001), and Dewey long ago made the same point about ethics (Dewey, 1922 [1988]: 74): however finely moralists articulate their favored principles, extensions will always be needed for application to new situations. An unremarked irony of the traditional perspective is that it effectively seeks to put an end to ethical life, by offering a system of principles so complete that conduct could properly be governed by consulting the definitive manual.

There is an alternative. Ethics can be conceived as an enterprise that grows out of the conditions of human life, that is permanently revised and expanded, and that is never finished. The primary task of the ethical theorist is to expose the character of this enterprise, the conditions that give rise to it, the nature of whatever progress it can achieve, and, building on these understandings, to consider the ways in which the ethical resources we have inherited might further be modified or elaborated. Our species has become what it is through its thorough immersion in the ethical project, and the principal goal of ethical theorizing is to learn how to continue that project more 'intelligently' (Dewey's term).

In what follows, I'll offer a condensed view of proposals I've developed in more detail elsewhere (Kitcher, 2011) — although it now seems to me that some central points can be highlighted without taking so firm a stand on controversial issues (the naturalism of this essay will thus be more ecumenical). First, I'll attempt to sketch the history of the ethical project. Second, I'll use that sketch to vindicate a concept of ethical progress. Third, I'll offer some suggestions about how the project might progressively continue. Finally, I'll try to answer my title question.

2. The evolution of ethical life

Ethical life grew out of the condition in which our ancestors lived. Ancestral hominids spent their lives in small groups (probably with 30 to 70 members) mixed by age and sex. Their ability to live in this way presupposed a capacity for responding to others: they could identify the wants and intentions of some of their fellows, and adjust their own behavior so that it helped their mates achieve their aims. Primatologists have documented the operation of that capacity in the actions of chimpanzees and bonobos (de Waal,

1996). Yet the capacity was — and is — limited. All too often our ancestors ignored the identifiable desires of their fellows, and their unresponsiveness led to social tension and conflict. The elaborate strategies of peacemaking, grooming huddles for example, are a striking expression of the need to re-pair a social fabric that is constantly in danger of being torn. Like chimps and bonobos, our pre-ethical hominid ancestors had enough responsiveness to the conspecifics around them to enable them to live together, but not enough to enable them to live together smoothly and well.

Other primates have evolved traits that we might count among the 'build-ing blocks' of ethics (de Waal, 2006). As Patricia Churchland has cogently argued, many primates share neural mechanisms that promote being together with conspecifics (Churchland, 2011). Besides the capacity for responsive-ness to others, chimpanzees and bonobos also display some tendencies for self-restraint: they can await 'their turn' for access to the '[nut]-cracking sta-tion', for example (de Waal, 2013: 149–150). In dyadic interactions, both chimpanzees and young children seem capable of recognizing the ways in which the situation is supposed to unfold, signaling a tacit awareness of norms (Tomasello, 2009). So we might credit our hominid ancestors with a bundle of psychological characteristics that were expressed in behavior we can retrospectively appraise as good or right. Furthermore, since the ho-minid line split off from that leading to our closest cousins, it is likely that purpose-built tendencies to cooperation have evolved, as for example with an enhanced tendency for older females to care for offspring of younger mem-bers of the group (Hrdy, 2009).

Full ethical life goes beyond these encouraging developments through the acquisition of an ability to conceptualize them differently, to see particular actions as good or as right, consciously to override particular desires, to dis-cuss potential patterns of conduct with others. For millennia, adult human beings have been able not only to 'do the right thing' but to appraise their conduct in this way, to worry about the rightness of what they intended to do, and to ask others for ethical advice. The tendencies present at earlier stages were made explicit, held up for scrutiny, and that's crucial to the ethical life as we live it. Very probably, this turn to explicit representation came about gradually, and it is quite unnecessary to decide where 'real ethics' begins. Nevertheless, it is important to understand the difference between the 'nice behavior' of our unselfconscious ancestors, and the reflective attitudes of the ethical project as it has figured throughout recorded history (and for far longer). Although a precursor to full ethical life was available prior to the acquisition of language, explicit consideration of what to do only came into being once our progenitors could speak to one another and to themselves. The appendices to codes of norms that appear in the earliest surviving documents make it plain that explicit representation of patterns for conduct has been part of human existence for tens of thousands of years, and a conservative estimate for the origins of ethical reflections would propose a 50 000 year history. If human language evolved much earlier, the ethical project, in its reflective form, could have been underway for considerably longer — and it is possible that the advantages of ethical discussion played an important role in the evolution of language.

An hypothesis: the tensions and uncertainties that attend the social lives of chimpanzees and bonobos continued to be present in hominid societies, even after whatever refinements of psychological traits occurred after the split that initiated the hominid line, and the elaboration of a reflective capacity for governing behavior is a response to the underlying problems that generate those tensions and uncertainties. The hypothesis rests on two ideas that are (I hope) plausible. First, even in the presence of an explicit system of elaborated norms, the difficulties posed by our incomplete responsiveness to others do not disappear entirely (to understate!). Second, the ability to foresee the damaging consequences of actions, combined with some capacity for self-restraint, is likely to mitigate the frequency with which limited responsiveness to others generates conflict and social tension; the ability to discuss and to agree on patterns for joint existence and joint action reaps further advantages of the same kind. To put the idea in its simplest terms, the facts of pre-hominid evolution placed our ancestors in a particular type of social environment, one they could handle but not handle well, and the achievement of reflective ethical life expanded already existent psychological tendencies to overcome the recurrent problems generated by the limitations of those tendencies.

Most of the history of the ethical project has been spent in small societies, in which at most fifty adult members have participated, on terms of rough equality (Lee, 1979; Knauft, 1991; Boehm, 1999), to work out how they should live together. From fifty thousand years before the present to the late Paleolithic, human beings engaged in numerous 'experiments of living', and the most successful elements of their practices were inherited by those who crafted the rules for the earliest cities, and were passed on to the ethical codes of today. Along this cultural evolutionary tree many important events occurred. Although the evidence is too scanty to support any definitive story, some points are relatively clear. About twenty thousand years ago, larger human groups began to assemble, at least temporarily. Almost certainly, this development required an expansion of the codes of the individual bands, so that protections assured to group members were extended, in some contexts, to outsiders; I suspect that these extensions are driven by the advantages of inter-group trade. Even in the earlier, smaller, bands, there may have been incipient division of labor, and this was probably amplified as groups grew larger. From division of labor emerged particular types of institution, and, with the advent of domestication and the transition to pastoralism, a concept of private property. Increased cooperation, particularly in stable partnerships, enhanced the repertory of human desires and feelings, eventually generating the concept of an enduring relationship in which each party attunes emotion, thought and action to the corresponding states in the other. Out of all this comes the richer conception of the good human life, already discernible in the most ancient texts we have.

The beginnings of the project were surely very crude. Our ancestors responded to the most salient problems, and, if we conceive of their situation as akin to those of contemporary chimpanzees or of contemporary hunter-gatherers, those would have stemmed from scarcity of resources and from tendencies to initiate violence. Ethics probably started with simple maxims, directives to share and prohibitions against unprovoked aggression. Motivation to conform to those precepts would have been grounded in fear: temptation to deviate was quelled by imagining the punishment to come (later, the original simple rules became part of a far more elaborate code, and the motivation to follow them drew on a broader array of emotions — solidarity, pride, shame, awe, as well as fear, but there is no moment at which some distinctive 'moral point of view' was attained; the idea of any such standpoint is a philosophical fiction).

All the groups in which ethical experimentation occurred faced an obvious problem: how to motivate conformity to the agreed-on code when the band-members are dispersed and conduct cannot be monitored? The ethnographic record reveals a prevalent solution (Westermarck, 1903: Chapter 50). Perhaps drawing on a prior conception of unseen beings, or perhaps introducing them for the job, our ancestors hypothesized an unobserved observer, a transcendent policeman, whose will is expressed in the ethical precepts of the group and who will punish those who violate them. Groups able to inculcate fear of the transcendent policeman would be likely to achieve higher rates of conformity, and, we can suppose, to reap corresponding social benefits. From this perspective, the hypothesis of the policeman is a helpful innovation.

Yet that hypothesis changes the character of ethical life. Once the deliverances of the code have been linked to the policeman's will, the process of revising or extending that code is subject to a new standard. Reaching agreement with one's fellows is no longer enough — the modifications should correspond to what the policeman wants. If someone in the group can claim special access to the policeman's will, that individual can pre-empt the previously democratic deliberation. The idea of an ethical expert is born, and ethical experts have the power to shape ethical life according to their own, possibly idiosyncratic, convictions.

3. Progress and ethical objectivity

Assuming that something like this narrative of the evolution of ethical life is correct, what bearing does it have on the questions about ethics that have provoked philosophers to construct their theories? One important issue concerns ethical objectivity, often presented in the form of asking whether it is possible to make sense of ethical truth. Denying any concept of ethical truth beyond that of social agreement appears to lead to an uncomfortable relativism, in which the precepts of each society count as 'true for them'. To the extent to which the agreements were arbitrary, the history of ethics would dissolve into a haphazard sequence of changes, without any direction. On the other hand, efforts to make sense of ethical truth typically seem either to invoke mysterious sources — divine beings whose utterances are the source of truth or abstract values that raise serious worries about how they might ever be apprehended — or to settle for the naturalistic surrogates (balance of pleasure over pain, tendency to spread human DNA, and the like) that have been so trenchantly criticized.

The recorded history of human ethical life does contain moments at which progress appears to me made, episodes in which our predecessors might be thought of as 'discovering hitherto unappreciated ethical truths'. Abandoning the view that chattel slavery is permissible looks like a progressive step: it is not easy to suppose that there is no important ethical difference between banning slavery and reintroducing it. Nevertheless, the character of the discovery made by the pioneering abolitionists remains mysterious, no matter how closely we probe the records of their thoughts and their lives. Normally discoveries consist in a change in the cognitive relations between the discoverer and some aspect of reality. But there is no apparent difference in relations to reality between the early abolitionists and the confident slave-owners who resisted their demands. If there are sources of ethical truth, they simply do not show up in the history of ethical change.

To think of a pre-existent property of wrongness, detected by those who first came to appreciate the wrongness of slavery, is a bad philosophical idea, of a piece with the conception of some antecedent subject-matter about which ethics is to provide a complete theory. Ethical objectivity requires a notion of ethical progress, but it does not need the thought that progress consists in the apprehension of truths already grounded in some aspect of reality. Instead of supposing progress to be towards a fixed goal, we should recognize the possibility of progress away from problematic situations. We make medical progress not by approximating some ideal of perfect health — there is no such ideal — but by overcoming the medical problems that beset people. The narrative of Section 2 makes this notion of progress vivid. Our remote ancestors made ethical progress by finding ways to overcome the limitations of responsiveness to others, a deep problem posed by our existence as social animals, whose symptoms were the recurrent tensions and uncertainties of hominid life.

Some people think that understanding progress in terms of problem-solving makes the concept subjective, and thus fails to escape the threat of relativism. They claim (echoing Berkeley) that to be a problem is to be felt as a problem. But this idea is mistaken. In many instances, we are unaware of the problems we face. Furthermore, as the comparison with medicine makes clear, problems that are felt often derive from objective features of the situation. It would be silly and insensitive to say of the cystic fibrosis patient in pulmonary crisis that her problem stems from her wish to breathe, or of the schizophrenic that his problem is grounded in a subjective wish that the tormenting voices should cease. These are not merely idiosyncratic desires, but reactions that almost any human being would have, if placed in the same situation. The problem is generated by the circumstances, even when those it besets have an intense wish for relief. In the particular case of hominid social life, the combination of the need to live among others and the limitations of responsiveness constitute the objective grounds of a problem. The ethical project begins in attempts to solve that problem (not fully conscious, of course, as in the early stages of human attempts to address medical problems). At the early stages, progress consists in achieving partial solutions.

Yet, like other problem-solving practices, ethics introduces new problems as it succeeds in tackling earlier difficulties. The eruption of conflicts around scarce resources was almost certainly a salient symptom of the problem of limited responsiveness, recognized at the beginning of the ethical project and addressed by introducing directives to share. Later, as the division of labor decreased the level of sacrifice demanded by increasing the supply of resources, human lives began to be more finely differentiated from one another, and the variations were intensified with the introduction of particular roles and institutions. Over many generations, ethical guidance of social practices has proliferated the possibilities for human living, equipping those who come later with a richer menu of options and of desires. One of the results of these developments is a tension between the idea of responding to all members of a society as equals and the idea of preserving and further expanding the varieties of human flourishing: egalitarians insist that the careers of the most fortunate are properly constrained by demanding for all the preconditions for worthwhile lives; their opponents emphasize ideals of human excellence, not to be compromised by instituting mediocrity for all.

The deepest difficulties in ethics result from situations in which problems conflict, when solutions to one set interfere with addressing others. So, while at the early stages of the ethical project, it is possible to speak unambiguously about ethical progress, in terms of the ur-problem of limited responsiveness, a complete concept of ethical progress presupposes some method for resolving problem-conflict. Not only are we committed to an evolving project, but also to evolving conceptions of progress and of method in ethics. There is no last word. Yet to acknowledge that fact should not tempt us into failing to take seriously the best approaches we can achieve, from what is — admittedly — a historically bound and limited perspective. For the same applies to the natural sciences as well: they have no last word either.

4. Going on

The grip of the standard framework for thinking about ethical theory — there is something antecedent to us and our ethical practices, and the task is to provide a completely systematic account of it — shows up in decisions about how to continue the ethical project. Faced with the incompleteness of the current code, revealed in ethical questions to which it cannot be uncontroversially applied, or with debate about particular features of that code, people have to work out how to extend or to revise what they have inherited (as Owen Flanagan pointed out to me, Alistair MacIntyre has been one of the few recent philosophers to be concerned with these issues; see MacIntyre (1988)). How can that be done? Since the dominant understanding of the ethical system throughout human history has been a spin-off from the idea of the transcendent policeman — the proper modification of the code should conform to the policeman's will — the appropriate method of discovery is one that fathoms the divine precepts. So people pray for guidance, or they listen to the advice of shamans, or of mystics, or of priests. It hardly matters that the reliability of these procedures is extremely dubious, or indeed that the whole idea of the divine will as the source of ethics has been problematic since Plato. It hardly matters because, within the standard framework, the secular alternatives are so thin and unconvincing.

Consider a standard practice of contemporary philosophy. Allegedly, you can find out answers to questions about particularly difficult ethical cases, and even about general ethical principles, by consulting your 'intuitions' about extremely schematic cases. "You find yourself beside a trolley track, on which five people are bound, and. . . " and so forth. By reflecting on these cases you are supposed to detect the sources of goodness and badness, rightness and wrongness. Unfortunately, few of those who have a taste for this sort of 'theorizing' offer a clear conception of what the sources are, and none have a convincing psychological story about how the recommended reflections might put you in touch with them (in fact, there are excellent reasons to think that our adapted psychological capacities are probably ill-suited to handle the cartoonish, and often bizarre, scenarios, dreamed up in this philosophical genre). By contrast, even though it is unlikely to be correct, the religious alternative does at least have a *story* about how prayer might yield correct ethical guidance.

Section 3 proposed that there is a coherent concept of ethical progress in terms of problem-solving. That concept can be applied to those moments in history at which ethical progress seems to occur: for example, the early abolitionists can be viewed as reacting against a particular way in which responsiveness to others was limited. To maintain the coherence of the concept, however, is not to claim that ethical progress is common in the history of our species, or that, when it occurs, that happens as the result of any clear insight. Most of the ethical progress about which we know seems to be made blindly, often when the voices of people whose aspirations have not previously been recognized become impossible to ignore (resistance at Stonewall began a series of ethically progressive steps that have since been guided in more reflective ways).

From the alternative perspective proposed here, the vision of ethics as an evolving project, always incomplete, the principal task of ethical theorizing is to provide tools that can be employed to make future ethical progress more frequent and less blind. To use Dewey's favorite term, the hope is to go on more 'intelligently'. Thinking about a more intelligent method for modifying our current ethical concepts and principles, or for developing our ethical resources to address open questions, might appropriately be informed by understanding what the ethical project is and how it has evolved. Thus the discussion of previous sections might guide our search for a method.

Although the introduction of the transcendent policeman was a clever device for increasing the probability that group members would conform to the agreed-on precepts, even when they were no longer under the scrutiny of their fellows, it distorted the ethical project. With the idea of an external standard came the issue of finding ways to gain access to it — generating the acceptance of religious teachers or (in some more recent traditions) private conversations with the deity. Part of the distorting idea is then preserved in the speculations of secular philosophy, which avoid the evident mistakes and uncertainties of the more vivid religious versions by offering pallid substitutes that, understandably, fail to convince many people that there is any secular alternative. Seeing the ethical project for what it is exposes the fact that the only resources we have ever had for inaugurating and continuing ethical life consist in conversations with others, and the chains of reasoning those conversations can initiate. Ethics begins with the problems of the human predicament — specifically with the problem of how intelligent animals with limited responsiveness to one another can live relatively easily with one another — and solutions to those problems may be achieved by blindly happening on something that improves the situation, or, perhaps, by our figuring out together how things might go better.

So I offer two proposals, one substantive and one methodological. The substantive suggestion is that, at our particular historical stage, when the development of human technology allows for the provision for all members of our species the preconditions for a worthwhile life — including not only the basic necessities, but also protection against disease and disability and access to education — all ethical discussion should accept, as a constraint, the task of providing for all the opportunity for a worthwhile life (and doing so on terms of rough equality). The methodological suggestion is that ethical decision-making should proceed as if it were the outcome of an ideal conversation, one in which the participants included representatives of all human points of view, in which each participant was required to modify her perspective through expunging all identifiable errors, and in which each was dedicated to discovering a solution with which all could live. These proposals embody an intention to return to the ethical project in its original, undistorted, form, and to scale it up to our present circumstances. Our early ethical ancestors began by attempting to secure certain basic things for all members of a small band: we have achieved a richer menu of desires and aspirations, and should attempt to enable all members of our species to have serious chances of reaching their goals. The ethical project began on terms of rough equality and focused on situations of direct experience: we should emulate the equal inclusion of all, import expertise where we can, and mimic the mutual engagement once achieved in 'the cool hour'.

These can only be proposals. For one of the leading implications of my account of the ethical project is that there are no ethical experts in the sense of authorities who can pronounce on ethical questions. (There are experts in a different sense, people who are well-equipped to facilitate conversation, to sharpen questions and distill the proposals made by others; philosophers may offer valuable service in this regard.) The collective process, in which perspectives are refined in light of factual knowledge and in awareness of the standpoints of others, has the last word.

It is worth emphasizing the gap between the ideal form of ethical discussion and the circumstances of actual conversation. As things actually stand, exchanges even incompletely representative of the full range of human perspectives would surely break down at an early stage, as participants insisted on the final validity of particular religious texts and traditions. Ideal conversation requires the expunging of falsehoods — and prominent among the falsehoods are claims that religions furnish ethical truths that cannot be challenged. Collective ethical discussion can no more tolerate perspectives that retain those falsehoods than it can allow participants to maintain the intrinsic value of short lives that end in agonizing pain. Consequently, there is an acute *political* problem in instituting conditions under which human ethical life can continue in an undistorted fashion, for the tendency to subordinate ethics to the illegitimate authority of (different) religions must first be eradicated (in the discussion in Erice, Ara Norenzayan raised some questions and challenges that have prompted me to make explicit the distinction between ethical theory and the conditions that would allow theory to be realized in practice).

5. So ... is a naturalized ethics possible?

Finally, I return to my title question, understood not in the political form just noted, but as an intellectual concern. As Section 1 already noted, attempts to naturalize ethics typically face two types of challenge: Do they offer the right ethical assessments? Can they avoid the naturalistic fallacy? I'll take these questions in order.

My version of naturalistic ethics does not directly yield judgments about the goodness of states of affairs or the rightness of actions. Instead it characterizes a procedure for arriving at such judgments. Imagine then that an ethical question is posed, the procedure is followed and an answer is delivered. On what basis might anyone claim that the answer is incorrect? Given my understanding of how the ethical project arose, how it evolved, and how it has been distorted through the invocation of individual authority, it is difficult to see how any rival answer could claim greater authority. The canvassing of opinions among mutually engaged interlocutors whose perspectives had been refined through expunging factual errors looks about as good as humans could ever do on any ethical issue. Surely this is superior to the dogmatic assertions of religious teachers or the 'intuitions' of individual philosophers, nurtured on abstract cases remote from the conditions of human life.

The first challenge must eventually rest on thinking that there is a better alternative to the method advocated in the previous section, a superior way of going on with the ethical project — and that turns out to be the deepest issue behind the second challenge. So let's turn to the charge that naturalism inevitably commits a fallacy.

As Richard Joyce has carefully argued (Joyce, 2006), there is no single thing counting as the naturalistic fallacy. There is a cluster of worries about the authority of the deliverances of a naturalistic ethics. The central issues can be focused by starting with Hume's old question about inferences from 'is' to 'ought'. You might try to apply that Humean worry against my version of naturalism (as people sometimes do) by asking how I can derive conclusions asserting that we should respond to others from facts, perhaps facts about the conditions of hominid life, perhaps facts about the subsequent unfolding of the ethical project. But this criticism rests on confusion. Why should anyone want or need that sort of derivation? Presumably, because the conclusions stand in need of justification. Yet derivations of that type would only be pertinent to justification if we were in the predicament of knowing lots of facts, and using those facts to justify normative conclusions. That predicament is not ours. People do not wander around the world, acquiring masses of factual knowledge until, one day, they defy Humean gravity by launching themselves into normative life through some inference of extraordinary levitative power. We are born into the ethical project, equipped from the start of our active lives with a mix of beliefs about facts and values. The justificatory problems that arise for actual human beings concern justifying changes in our ethical judgments.

The methodological proposal of the last section specifies a standard of progress: a change in ethical practice would be progressive just in case it would be approved in an ideal discussion. Actual ethical agents make justified changes in their ethical beliefs when they do the best they can to simulate such discussions. In practice, this is a matter of thinking about things from different points of view, trying to get rid of factual errors, vividly identifying with other perspectives — an activity often stimulated by actual conversation, by learning pertinent facts, and by the imaginative provocations of literature and the arts (these serve far better, in my view, than the cartoon cases supposed to evoke philosophical 'intuitions'). These are the ways we actually have for 'bridging the Humean gap'. We could surely improve them, if our conversations ranged over a broader class of interlocutors, if our factual knowledge were greater, and our sympathies could be more constantly aroused.

My account of justified ethical change rests on two assumptions. First, it supposes that people should continue the ethical project, that they cannot simply give up the practice of regulating their lives. Second, it takes the proper continuation of the ethical project to be constrained by the original problem that sparked it, the problem of limited responsiveness. The method of Section 4 rests on giving priority to a search for agreement with all those with whom we causally interact (in some contexts, the entire species, including our descendants).

The first assumption is easily defended. We have no alternative but to continue the ethical project. To abandon it would be to return us to the predicament of our hominid ancestors — a predicament even more problematic for us than it was for them. The second assumption raises the same question that emerged from the earlier challenge (about whether my naturalism delivers the right verdicts): Is there an alternative, superior, proposal for continuing our ethical life?

Earlier, I claimed that problem conflict is the source of the difficulties of ethics. My preferred method takes the ur-problem — limited responsiveness — to remain important. Perhaps, though, if we saw some different problem as taking priority, that would vindicate a different method for continuing the ethical project. The obvious suggestion is that the evolution of the project has transformed human life, yielding dramatically different possibilities of human flourishing, and, once these are available, the central problem becomes that of allowing those who are capable of realizing them to do so. We should forget about the mediocre aspirations of the many ('the herd') and promote the glorious lives of the few ('the free spirits'). Nietzsche presents the deepest challenge to the democratic (egalitarian) version of naturalistic ethics I have commended.

Suppose, however, we took seriously the thought of promoting the highest forms of which human lives are capable, even if it treated many human lives as unworthy of serious support. How would we do this? Babies do not emerge into the world with labels of their special potential attached. In practice, the best we can do to encourage 'human flourishing at its finest' is to nurture all human beings, providing all young people with opportunities for a worthwhile life. A Nietzschean program is perforce drawn to the egalitarian ideal of Section 4.

Yet there is a more basic reason for entertaining both proposals of Section 4. Although the scope of our interactions with other human beings has been radically extended, the predicament from which our ancestors began persists — indeed, because of the distances that separate agents from those affected by their actions, the problems of limited responsiveness are exacerbated. My history of the ethical project shows both the omnipresence of this fundamental human difficulty and the distortions that have turned our ethical lives away from confronting it directly. Understanding the genealogy of ethics provides motivation for trying to scale up the original form of ethical life.

Philosophers impressed with Humean worries about deriving 'ought' from 'is' will surely protest that history cannot provide directions for continuing a project. "What form of cogent inference leads from the narrative of the ethical project to the conclusion that the problem of limited responsiveness remains of primary concern?" The question cannot be answered by offering some extension of formal logic — I freely confess that I do not know how to formalize the idea of learning from history. Yet, as we know from the historical research initiated by Thomas Kuhn (Kuhn, 1962), there are major episodes in the growth of the sciences that are not resolved on the basis of arguments admitting familiar styles of formalization. Judgment pervades. With respect to important scientific decisions, the best that can be done is to present the considerations that led to resolution, sometimes comparing them with other examples, so that their reasonableness becomes apparent.

So too with my naturalistic ethics. The proposals I have made are generated in similar ways to many that are familiar from our individual ethical lives. We discover that what we have said or done arouses protests in someone affected by our actions, and we resolve to be more sensitive, more responsive, in future. By the same token, the genealogy of ethics reveals the forms of our past collective blindness, pointing up the root cause. Limited responsiveness is the secular counterpart of original sin — and we should recognize it as such.

Acknowledgements

I am grateful to the participants in the Erice workshop for comments and questions that have helped me to develop my talk into this essay. Thanks also to Owen Flanagan and Patricia Churchland for valuable comments and encouragement.

References

Bentham, J. (1988). Principles of morals and legislation. — Prometheus Books, Amherst, NY.
Boehm, C. (1999). Hierarchy in the forest. — Harvard University Press, Cambridge, MA.
Cartwright, N. (1999). The dappled world. — Cambridge University Press, Cambridge.
Churchland, P. (2011). Brain trust. — Princeton University Press, Princeton, NJ.
de Waal, F.B.M. (1996). Good natured. — Harvard University Press, Cambridge, MA.
de Waal, F.B.M. (2006). Primates and philosophers. — Princeton University Press, Princeton, NJ.

de Waal, F.B.M. (2013). The bonobo and the atheist. — W.W. Norton, New York, NY.

Dewey, J. (1922 [1988]). Human nature and conduct. — University of Southern Illinois, Carbondale, IL.

Dupre, J. (1993). The disorder of things. — Harvard University Press, Cambridge, MA.

Hrdy, S. (2009). Mothers and others. — Harvard University Press, Cambridge, MA.

Joyce, R. (2006). The evolution of morality. — MIT Press, Cambridge, MA.

Kitcher, P. (2001). Science, truth, and democracy. — Oxford University Press, New York, NY.

Kitcher, P. (2011). The ethical project. — Harvard University Press, Cambridge, MA.

Knauft, B.M. (1991). Violence and sociality in human evolution. — Curr. Anthropol. 32: 391-428.

Kuhn, T.S. (1962). The structure of scientific revolutions. — University of Chicago Press, Chicago, IL.

Lee, R. (1979). The !Kung San. — Cambridge University Press, Cambridge.

MacIntyre, A. (1988). Whose justice? Which rationality? — University of Notre Dame Press, Notre Dame, IN.

Parfit, D. (2011). On what matters (2 volumes). — Oxford University Press, Oxford.

Ruse, M. & Wilson, E.O. (1986). Moral philosophy as applied science. — Philosophy 61: 173-192.

Tomasello, M. (2009). Why we cooperate. — MIT Press, Cambridge, MA.

Westermarck, E. (1903). Origin and development of the moral ideas. — MacMillan, London.

[When citing this chapter, refer to Behaviour 151 (2014) 261–278]

The origins of moral judgment

Richard Joyce [*]

Department of Philosophy, Victoria University of Wellington,
Murphy Building, Kelburn Parade, Wellington 6140, New Zealand
[*]Author's e-mail address: richard.joyce@vuw.ac.nz

Accepted 15 September 2013; published online 27 November 2013

Abstract

Is human morality a biological adaptation? And, if so, should this fact have any substantial impact on the ethical inquiry of how we should live our lives? In this paper I will address both these questions, though will not attempt definitively to answer either. Regarding the former, my goal is to clarify the question and identify some serious challenges that arise for any attempt to settle the matter one way or the other. Regarding the latter, my ambitions here are restricted to some brief critical comments on one recent attempt to answer the question in the affirmative.

Keywords

evolution of morality, evolutionary ethics, adaptation versus spandrel, debunking arguments.

1. Introduction

Let us start with Darwin:

> I fully subscribe to the judgment of those writers who maintain that of all the differences between man and the lower animals, the moral sense or conscience is by far the most important. … [A]ny animal whatever, endowed with well-marked social instincts, the parental and filial affections being here included, would inevitably acquire a moral sense or conscience, as soon as its intellectual powers had become as well, or nearly as well developed, as in man. (Darwin, 1879/2004: 120–121)

There are several features of this passage worth highlighting. First, the trait that is under discussion is described as 'the moral sense or conscience', which, it seems safe to claim, is a faculty that produces moral judgments.

Darwin is not here wondering whether being morally good is the product of evolution, but rather whether the capacity to make self-directed moral judgments is the product of evolution. A moment's reflection on the myriad of ways in which morally appalling behavior may be motivated by a sense of moral duty should suffice to illuminate the distinction.

The second conspicuous feature of the passage is that Darwin sees the moral sense as emerging (inevitably) from other traits: 'social instincts' combined with 'intellectual powers'. The latter powers he goes on to mention are memory, language, and habit. This raises the possibility that Darwin does not see the moral sense as a discrete psychological adaptation but rather as a byproduct of other evolved traits. In fact, he appears wisely to steer clear of adjudicating on this matter. When focused on the social instincts generally (rather than the moral sense in particular), he writes that "it is... impossible to decide in many cases whether certain social instincts have been acquired through natural selection, or are the indirect result of other instincts and faculties" (1879/2004: 130).

Contemporary debate among philosophers (in particular) over whether the human moral sense is an adaptation has not always been so cautious. Several recent authors have developed arguments to the conclusion that human moral judgment is not a discrete adaptation but rather a byproduct of other psychological traits (Nichols, 2005; Prinz, 2008; Ayala, 2010; Machery & Mallon, 2010). Let us call these people 'spandrel theorists' about morality. Others, myself included, have advocated the view that the human moral sense is a biological adaptation (Alexander, 1987; Irons, 1996; Krebs, 2005; Dwyer, 2006; Joyce, 2006; Mikhail, 2011). We'll call these people 'moral nativists'. My first substantive goal in this paper is to reveal how difficult it is to resolve this matter.

2. Adaptations versus spandrels

The spandrel theorist proceeds by offering 'non-moral ingredients' — themselves quite possibly adaptations — which are sufficient to explain the emergence of moral judgment. We have seen Darwin mention such things as language use, social instincts, and memory. Francisco Ayala emphasizes "(i) the ability to anticipate the consequences of one's own actions; (ii) the ability to make value judgments; and (iii) the ability to choose between alternative courses of action" (2010: 9015). Jesse Prinz (2008, 2014) considers such non-moral ingredients as meta-emotions, perspective taking, and the capacity for abstraction. Here I will take as my exemplar the view of Shaun Nichols (2005), but the general point I shall make could be leveled at any of the aforementioned (and, indeed, against any spandrel theorist).

The two non-moral ingredients that Nichols focuses on are a capacity to use non-hypothetical imperatives[1] and an affective mechanism that responds to others' suffering. He writes that:

> ...both of the mechanisms that I've suggested contribute to moral judgment might well be adaptations. However, it is distinctly less plausible that the capacity for core moral judgment itself is an adaptation. It's more likely that core moral judgment emerges as a kind of byproduct of (*inter alia*) the innate affective and innate rule comprehension mechanisms. (2005: 369)

An obvious way of critically assessing Nichols' claim would be to question whether these two mechanisms, working in tandem, really are sufficient to explain moral judgment (for the sake of simplicity I'm ignoring Nichols' sensible 'inter alia' in the previous quote). This would involve describing the two mechanisms highlighted by Nichols in much more detail, searching for empirical evidence (e.g., can an individual have one of these mechanisms impaired and yet still make moral judgments?), and so forth. But the question I want to ask is much more general: What determines whether a trait (i) is a byproduct of other mechanisms x, y, and z, or (ii) is an adaptation dependent upon pre-adaptational sub-mechanisms x, y and z? Answering this question in the abstract is fairly straightforward, but having a procedure for empirically determining whether a trait is one or the other is considerably more difficult. Let me explain.

---

[1] A hypothetical imperative (e.g., 'Go to bed now') recommends that the addressee pursue a certain means in order to achieve one of his/her ends (e.g., to get a good night's sleep). If it turns out that s/he lacks that end, then the imperative is withdrawn. A non-hypothetical imperative demands an action irrespective of the addressee's ends. For example, the imperative 'Do not slaughter innocents' is not withdrawn upon discovery that the addressee loves slaughtering innocents, will not get caught, and does not give a fig for morality. Moral imperatives are a subset of non-hypothetical imperatives. Non-moral non-hypothetical imperatives include etiquette, road regulations, rules of games and sports, and the norms of institutions generally.

No psychological faculty for producing a species of judgment is going to exist as a monolithic entity that takes inputs and magically produces outputs; all such faculties will depend on the operation of numerous psychological sub-mechanisms, which in turn depend on sub-sub-mechanisms, etc. Suppose that Nichols is correct that the two mechanisms he highlights are indeed sufficient to explain the phenomenon of moral judgment. One interpretation — the one Nichols favors — is that the capacity for moral judgment is a byproduct of the operation of these two mechanisms. But a second hypothesis is always available: that the capacity for moral judgment is a distinct adaptation of which these are two sub-mechanisms. The second hypothesis is true if (and only if) the manner in which these two mechanisms interact has been at all modified by natural selection because their interaction has some impact on reproductive fitness. Let us suppose first of all that these two mechanisms evolved for their own evolutionary purposes. But in certain circumstances they interacted, in such a way that the trait of moral judgment emerged as a byproduct. Suppose further, however, that this new trait (moral judgment) had some reproductive relevance, such that the process of natural selection began to 'tinker' — perhaps strengthening the interaction of the two mechanisms in some circumstances, dampening it in others. If this has occurred, then the capacity for moral judgment is no longer a mere 'byproduct' but rather an adaptation in its own right (of course, one can still maintain that it originally appeared as a byproduct, but this is true of virtually everything that counts as an adaptation; see Dennett, 1995: 281).

In sum, spandrel theorists about morality seem to think that it suffices to establish their view if they offer non-moral ingredients adequate to account for moral judgment. But the consideration just raised indicates that this matter is not so straightforward, for any spandrel hypothesis can be interpreted instead as a description of the sub-mechanisms of the nativist moral sense (and if the ingredients mentioned are indeed adequate to explain moral judgment, then so much the better for the resulting nativist hypothesis).

But how would one distinguish empirically between these two hypotheses? The difference between an adaptation and a byproduct cannot be discerned by consulting intrinsic features of the organism, no matter in what detail. Consider Stephen Jay Gould's architectural analogy that originally provided the term 'spandrel' (Gould & Lewontin, 1979). Renaissance architects faced the design challenge of mounting a dome upon a circle of arches; when this is accomplished, the spaces between the arches and dome produce roughly triangular areas of wall: spandrels. These areas of wall are not design features — they are byproducts of the design features. Yet one could not discern this by examining the intrinsic structural features of the building; one must know something about the purposes of the architects. It is, after all, conceivable that an architect may have a direct interest in creating spandrels, in which case the dome and arches would be byproducts. The resulting church would be intrinsically indistinguishable from the ordinary church for which the spandrels are byproducts.

In the same way, in order to know whether a trait is an adaptation as opposed to a byproduct one must understand something of the intentions of the architect — in this case, the forces of natural selection that existed during the period of the trait's emergence. Lacking, as we usually do, concrete evidence of the subtle evolutionary pressures operating upon our ancestors, our epistemic access to this information will always depend to some extent on intelligent inference. Consider, for example, Nichols' contention that the capacity to use non-hypothetical imperatives is an adaptation whereas the capacity to use moral imperatives is a byproduct. An alternative view is that the capacity to use moral imperatives is the adaptation while the more general capacity to use non-hypothetical imperatives is the byproduct. One could not decide between these hypotheses simply by examining the human organism; rather, the decision would have to involve comparing the plausibility of two conjectural hypotheses. On the one hand, one might hypothesize that the ancestral environment contained adaptive problems for which the specific capacity to use moral judgments would be a reasonable solution. Alternatively, one might hypothesize that the ancestral environment contained adaptive problems for which the specific capacity to use non-hypothetical imperatives would be a reasonable solution. In either case, the adaptive problems would need to be described in a manner supported by available evidence. To the extent that the former hypothesis turned out to be more plausible than the latter, moral nativism would be supported. But if the latter were more plausible than the former, then support would be provided for the spandrel view. A troubling possibility, of course, is that we may very well find ourselves lacking solid ground for favoring either kind of hypothesis over the other, in which case we'd lack ground for claiming with confidence which trait is the adaptation and which the byproduct. One can see now, perhaps, the wisdom of Darwin's quietism on this matter.

3. What is the trait under investigation?

I have been outlining one way in which the dispute between the moral nativist and the spandrel theorist is likely to run aground. However, it might reasonably be responded that this problem is of little consequence, since the contrast that is of greater theoretical interest is whether the capacity to make moral judgments is the product of evolutionary forces (whether an adaptation or a byproduct) or is an acquired ability. Frans de Waal calls the latter position 'veneer theory': the view that morality, along with cooperative and altruistic tendencies in general, is "a cultural overlay, a thin veneer hiding an otherwise selfish and brutish nature" (de Waal, 2006: 6). I doubt that many people nowadays endorse the veneer theory; that humans have been designed by natural selection to be gregarious and cooperative seems beyond reasonable doubt. The devil lies in the details of *how* we are gregarious and cooperative. Note that declaring that we are by nature gregarious and cooperative is not to declare in favor of moral nativism, for it remains entirely possible that our social nature consists of biologically entrenched tendencies toward altruism, sympathy, love, and so forth, while the capacity to make moral judgments is an acquired and relatively recent cultural characteristic.

This observation, however, focuses attention on the knotty question that lies at the heart of these debates: What is a moral judgment? There is little to be gained in arguing over whether a trait is an adaptation or a spandrel, innate or acquired, if we do not have a firm handle on the nature of the trait under investigation. It is a great inconvenience to these debates that the concept *moral judgment* is a slippery and highly contested idea even among those who are supposed to be experts on the topic — namely, metaethicists.

In order to approach this problem, let us pause to compare chimpanzee sociality with human sociality. De Waal has often claimed that chimpanzee life contains some of the 'building blocks' of morality (1992, 2006). He focuses on such things as reciprocity, consolation behavior, inequity aversion, empathy, and the following of rules of conduct reinforced by others. At the same time, de Waal is positive that chimpanzees do not make moral judgments (1996: 209; see also Boehm, 2012: 113–131). This raises the question of what additional building blocks need be added, or how the building blocks need be rearranged, in order to create something deserving of the name 'a moral sense'. The fact that the answer is not at all clear problematizes the whole dialectic concerning the evolution of morality. In what follows I will attempt to say something useful on the matter.

A striking feature of the chimpanzee building blocks is that they seem to require emotional arousal. A deviation from a social rule in chimpanzee society receives a negative response only because those giving the response get angry. Consolation behavior is provided only by those in whom sympathy has been stirred. A reciprocal act (grooming behavior, say) occurs because the reciprocator feels friendly and caring toward the recipient (or, perhaps, feels fearful of the reprisal that non-reciprocation might bring). What chimpanzees seem to lack is a psychological apparatus that could motivate such behaviors in the absence of emotional arousal. In humans, by contrast, a deviation from a social rule might receive a negative response because those giving the response judge that it is deserved; consolation behavior might be provided by those who considers it right to do so; reciprocation might be offered because one judges oneself to have a duty to repay a debt; and so forth.

There are many who would claim that because the prominent 'building blocks of morality' seem to be affective phenomena, the fully-fledged moral faculty must also be an affective mechanism.[2] One might argue, for example, that what humans have (which other primates lack) is the capacity to have meta-conations. Perhaps if an individual not only dislikes a certain behavior, but likes the fact that she dislikes it (and perhaps also dislikes anyone who fails to dislike it, and likes anyone who does dislike it) then we may speak of her 'morally disapproving' of the behavior. Perhaps if one's dislike of another's action prompts not only anger, but a disposition to feel anger at those who do not also feel anger at the action, then we may speak of one's judging that the anger is merited (see Blackburn, 1998: 9–13; Prinz, 2007: 113–115).

I find this line of reasoning unpersuasive. The building blocks of morality found in chimpanzees (and, by presumption, our ancestors) may well be affective phenomena, but it is entirely possible that the crucial modification of these building blocks in the human lineage was the addition of certain

---

[2] Here I am using the terms 'affective', 'noncognitive' and 'conative' synonymously. I am not shunning the term 'emotional', but am treating it with care, for emotions — at least many of them — are mixtures of affective and cognitive components (for this reason, I do not consider Christopher Boehm's claim that the internalization of norms requires that one "connect with these rules *emotionally*" (2012: 114) to be necessarily at odds with the cognitivist line I push in this paper).

cognitive aptitudes. After all, generally speaking, the explosion of cognitive abilities is surely the most striking aspect of recent human evolution. Moreover, it is far from obvious, just on conceptual grounds, that one can really build a moral judgment from these affective ingredients alone. The natural way of assessing the claim is to examine potential counterexamples, of which there are two types. First, can we imagine these noncognitive capacities being deployed without a moral judgment occurring? Second, can we imagine a moral judgment occurring without these noncognitive capacities being deployed? I'm inclined to think that the answer to both questions is 'Yes'.

Suppose I am strolling among a group of normally docile animals when one bites me aggressively. Naturally, I dislike this; perhaps I smack the animal on the nose in order to make it release me. Perhaps, moreover, I judge that it's important that these animals do not form aggressive habits (maybe my children often play in their vicinity), so I would wish to see others smack the animal if bitten. Perhaps I go so far as to dislike anyone who would not smack the animal if bitten. Yet these emotions and meta-emotions do not appear to amount to a moral judgment of the animal's behavior. It does not seem that I judge that the animal deserves to be smacked; indeed, I do not treat the animal's behavior as a transgression at all. I do not disapprove of its aggressive behavior; I simply dislike it in an elaborate way.

The reason we do not make moral judgments concerning animals is because they lack a certain kind of agency that we think of as a prerequisite for moral assessment (it does not matter to our current purposes what the nature of this agency is). Taking this into account, one might respond that the emotions that form the basis of moral judgment are a kind that can be coherently deployed only toward creatures that fulfill these criteria of agency. The 'dislike' felt toward a violent animal just is not the right sort of affective state to begin with (the response goes); perhaps talk of 'disapproval' would be more apt than talk of 'dislike'.

The problem with this response is that disapproval is not a mere noncognitive response; it is a mental state permeated with conceptual content. Disapproval requires a concomitant judgment that the object of assessment has transgressed in a manner that warrants some sort of punitive response (if only treating with coolness). One therefore cannot appeal to disapproval as the basic noncognitive state to explain meriting (for example). The same problem would emerge if one tried to account for moral judgment in terms of the emotion of guilt — for this is an emotion with conceptually rich components (see Joyce, 2006: 101–104). I therefore doubt that one can build moral judgments out of affective phenomena alone.

Not only are purely noncognitive building blocks insufficient for moral judgment, but they appear to be unnecessary. Consider a moral judgment voiced in circumstances of emotional fatigue. Perhaps one has just been exposed to a sequence of similar moral scenarios and one's capacity for emotional arousal has ebbed (maybe one is ticking the hundredth box on a psychology experiment designed to ascertain subjects' moral intuitions on a range of cases). Or perhaps one is simply distracted. All too often those who claim that emotional arousal is necessary for moral judgment focus on extreme cases: our disgust at pedophilia, our horror at the thought of the trains discharging their passengers at Auschwitz. Mundane moral judgments — like thinking that the gold medalist deserved her win, or that a person's ownership of his shoes grants him certain rights to that footwear — do not get a look in. One can claim, of course, that even for these mundane cases emotional arousal is *possible* (imagine someone having his shoes stolen; picture his outrage; visualize his suffering as he walks home barefoot through the snow), but emotional arousal to *anything* is possible.

This is one problem with Prinz's view that even if someone making a moral judgment is not emotionally aroused he or she is at least disposed to become emotionally aroused (2007: 84ff). Even if one could specify precisely what kind of emotion is relevant, there is simply no such thing as the disposition to have that emotion (occurrently) *period*; it must be a disposition to have that emotion (occurrently) in such-and-such circumstances. But while one may identify circumstances under which an individual might become emotionally aroused at the thought of someone's enjoying rights over his own shoes, so too one may think of circumstances under which an individual might become emotionally aroused at the thought that gold has atomic number 79 (or any other matter). It may be possible to find a principled distinction between such cases, but to my knowledge none has ever been articulated.

Highlighting the cognitive achievements inherent in moral judgment is not intended to exclude the affective components. As we have seen, affective mechanisms were probably central to the emergence of moral judgment — at least as pre-adaptations — and all the evidence indicates that emotions continue to play a central role in human moral life (see Haidt, 2001; Greene & Haidt, 2002; Wheatley & Haidt, 2005; Valdesolo & DeSteno, 2006; Small & Lerner, 2008; Horberg et al., 2011). None of this, however, undermines the hypothesis that certain cognitive capacities are necessary for moral judgment, and that these capacities were the key development — the crucial additional building blocks — in the emergence of human morality.

The cognitive capacities I have in mind might be described as those necessary for the 'moralization' of affective states. Consider the elaborate cluster of conations and meta-conations described earlier, which I doubted were sufficient for a moral judgment. What the cluster seemed unable to account for were ideas like disapproval, transgression and merited reaction (i.e., desert). Without these, the fully-blown moral conceptions of obligation, prohibition (and thus permission³) are unavailable. Without the concept of obligation, there is no possibility of judging anyone to have a right, and without rights there can be no idea of ownership (only the idea of possession).

The chimpanzee brain lacks the mechanisms necessary to access this conceptual framework probably as surely as the human brain lacks the mechanisms for navigating the world using echo-location. Even if we could ramp up the chimpanzee's capacity for meta-conations (allowing them, say, the capacity to get angry at those who do not get angry at anyone who fails to get angry at someone who does so-and-so), we still would not thereby grant them the capability for judging a punitive response to be deserved. Nor would we grant them this capability if we could boost their abilities to discriminate factual data in their environment (allowing them, say, the capacity to infer that if X desires Y's welfare, and X believes that Z will get angry at Y if Y performs action φ, then X will want Y to refrain from φ-ing). It cannot be the mere 'abstract' quality of moral concepts that places them beyond the chimpanzee's grasp, for in other ways chimpanzees wield abstract concepts smoothly.⁴ De Waal rightly claims that humans have a greater capacity to internalize norms than other primates (Flack & de Waal, 2001: 23; see also

---

³ If one lacks the concepts of obligation and prohibition, then one lacks the concept of permission. Contra Camus' claim that "if we can assert no value whatsoever, everything is permissible" (1951), if there are no moral values then *nothing* is morally permissible.

⁴ Consider a chimpanzee's postponing a vengeful act against a rival until a good opportunity arises. Perhaps we grant it deliberations about plans it will execute 'later' — but *later* is an abstract concept. Or consider the way that chimpanzees can play 'spot-the-odd-one-out'-type games (Garcha & Ettlinger, 1979). *Sameness* and *difference* are abstract concepts.

Boehm, 2012: 113–131), but the puzzle remains: What mechanisms does a brain need in order to have the capacity to internalize a norm? It is natural to answer by saying something about the fear of punishment becoming assimilated, such that the individual self-regulates behavior by administering his/her own emotional punishment system. But the puzzle reiterates. To fear punishment is not to have internalized a norm (since one can fear punishment for a crime that one does not believe really is a crime); for internalization, one must believe that punishment would be *merited* and thus be disposed to dispense a kind of punitive self-reproach to oneself even in the absence of witnesses. But what accounts for this concept of 'meriting'? Again I would answer that it is challenging to see how a creature could form such a thought using only purely conative and more general data-processing mechanisms (no matter how elaborate). I propose that norm internalization requires cognitive resources dedicated to normative thinking in particular.

The suggested hypothesis is that the human brain comes prepared to produce normative cognitions in a similar way that it comes prepared to encounter faces, other minds, and linguistic stimuli. This is not to say that it comes prepared for any particular normative system — that is, one with a particular normative content. The conspicuous phenomenon of moral disagreement demonstrates that moral content is learned and to some extent flexible, in the same way that the abundance of natural languages demonstrates that languages are learned and to some extent flexible. And to restate an earlier point: The hypothesis that the human brain comes prepared for normative thinking is a more general proposition than the moral nativist hypothesis. Perhaps Nichols is correct that non-hypothetical normative thinking is an adaptation while specifically moral thinking is a spin-off capacity. Or perhaps it is the other way round. Deciding whether something is an adaptation involves a large dose of inference and speculation concerning what we suppose were the relevant adaptive problems placing pressure upon our ancestors in the distant past.

Insisting on the cognitive components of moral judgment still leaves much undecided about the exact nature of these judgments. Some have argued, for example, that one characteristic of moral judgments is a particular kind of practical authority: moral rules (unlike those of most other normative systems) are those with which one must comply whether one likes it or not. Others have doubted this, allowing that a person with sufficiently aberrant goals and desires (and appropriately situated) may well have no reason to care about moral imperatives.[5] The cognitive quality of moral judgment is consistent with either view; it is silent of the subject. A disquieting possibility is that the notion of moral judgment is in fact not as determinate on this matter (or on other matters) as we generally presuppose. Perhaps there is simply no fact of the matter as to whether moral rules have or lack this authoritative quality. Certainly people seem to generally imbue their moral prescriptions with this kind of strong authority, so maybe having a theory that provides this authority is a theoretical desideratum. But perhaps this authority is not an indispensable component of morality; maybe if we can make no sense of this authority and have to settle for a normative system lacking it, the system would still deserve the name 'morality'. One way of diagnosing this situation would be to say that strictly speaking morality has this authoritative quality, but loosely speaking it need not.

Something similar has been said about language by Marc Hauser, Noam Chomsky, and W. Tecumseh Fitch, who argue that one can speak of language in a broad sense or a narrow sense (2002). The former consists of linguistic capacities that we share with other animals, whereas the latter includes the uniquely human trait of linguistic recursion. There is no answer to the question of which idea captures what is 'really' language; our vernacular concept of language is simply not so fine-grained as to license one answer while excluding the other. Faced with the query of whether vervet monkeys, say, have a language, the only sensible answer is 'In one sense yes and in one sense no'.

The same may be true of morality. The vernacular notion of a moral judgment may simply be indeterminate in various respects, allowing of a variety of precisifications, with no particular one commanding acceptance. This raises the possibility that the capacity to make moral judgments construed in one sense may be an adaptation, while the capacity to make moral judgments construed in another (equally legitimate) sense is not. One might even go so far as to say that chimpanzees satisfy the criteria for making moral judgments *very loosely construed* — though I would urge against liberality taken so far. A less excessive and not implausible possibility is that on some

---

[5] Philosophers who advocate the thesis that moral prescriptions enjoy some kind of special authority include Immanuel Kant, J.L. Mackie, Michael Smith, and Christine Korsgaard. Those who allow the possibility that one may have no reason to act morally include David Hume, Philippa Foot, David Brink, and Peter Railton.

broad construal of what a moral judgment is, the capacity to make them is a deeply entrenched part of evolved human psychology, while on a stricter construal the capacity is a recent cultural overlay: a veneer.[6]

4. Implications of cognitivism

Whether on any reasonable precisification of the concept *moral judgment* the cognitive element is necessary is something on which I will not attempt to adjudicate (though earlier arguments reveal my inclination to think so). Certainly I maintain that this element is necessary at least for moral judgments strictly construed. I will close by considering some of the implications of moral judgments being cognitive in nature.

To claim that moral judgments essentially involve a cognitive component is basically to claim that they essentially involve beliefs. For example, if one holds (as one should) that a judgment that a punitive response is deserved must involve something more than just elaborate conative attitudes, then one holds that it involves (possibly inter alia) *the belief* that the punitive response is deserved. Once beliefs are in the picture, then certain distinctive ways of assessing moral judgments must be permitted, meaning that human morality can be interrogated in ways that, say, chimpanzee social systems cannot be. A chimpanzee group may enforce a rule that is in fact practically sub-optimal; so too may a human group. An individual chimpanzee may become affectively aroused at another in a way that harms its own interests (or furthers its own interests); so too may a human individual. But the fact that the human moral faculty involves normative beliefs means that human moral judgments can be evaluated in additional ways for which evaluating the chimpanzee response would make no sense. Beliefs can be assessed for truth or falsity in a way that purely noncognitive states cannot be. Beliefs can be assessed for justification or non-justification in a way that purely noncognitive states cannot be (this is not to claim that all talk of justification is misplaced for noncognitive attitudes, but that it must be of a very different type[7]). Therefore, a human moral response may be probed with the questions

---

[6] This, clearly, would not be what de Waal means by 'veneer theory', since, on the view just described, morality (strictly construed) would be a veneer over a core of social and altruistic tendencies, not (as de Waal envisages) over a core of nasty asocial selfishness.

[7] A basic distinction here is between instrumental justification and epistemic justification. Something is instrumentally justified if it furthers one's ends. Mary's belief that the famine

'Is it true?' and 'Is it justified?'. And if one can do this for a token judgment, there seems nothing to stop one posing these questions on a grand philosophical scale: inquiring of human moral judgments in general 'Are they true?' and 'Are they justified?'.

Some may say that asking these epistemological questions of morality is somehow off the mark — that the more important question regarding human morality *is* the one that can also be asked of chimpanzee social regulation: namely, 'Does it work?' I find that I have nothing to say about which kind of question is more urgent or more interesting; it's a matter of what one's theoretical concerns are. I do think, however, that the epistemological questions can be legitimately asked of any belief, and it is the job of the metaethicist to press these questions hard regarding moral beliefs.

My approach to these matters puts me somewhat at odds with that of Philip Kitcher (2011, 2014). Kitcher sees moral judgment as having emerged for a purpose, allowing one to speak of its fulfilling its function well or poorly. This in turn allows one to make sense of moral progress, but not in the manner of scientific progress — that is, the attainment of improving approximations of the truth — but in the manner of refining a tool to better accomplish its task. Moral truth, for Kitcher, can enter the picture later: defined derivatively from the account of moral progress, not vice versa.

Kitcher and I agree that a 'moralization' of affective attitudes occurred at some point in our ancestry. In this paper I have advocated the view that what allowed this moralization were new building blocks of a cognitive nature: essentially, beliefs about behaviors being forbidden, punishments being just, and so forth. Instead of their large-scale cooperative projects being at the mercy of capricious conative states, our ancestors became able to think of cooperation (in certain circumstances) as absolutely required, of defection meriting penalty, etc., which supported a more robust motivation to participate. I'm inclined to think that Kitcher is correct in holding that the purpose of morality is, broadly, to augment social cohesion, but I would place more focus on how moral thinking accomplishes this end: by providing people

---

in Africa is really not so bad may be instrumentally justified (for her) if her knowing the truth would cast her into a depression. A belief is epistemically justified if it is formed in a way that is sensitive to the evidence. Mary's belief that the famine in Africa is not so bad, though it makes her happier, is epistemically unjustified if she has been exposed to sufficient evidence of its falsehood (which she ignores). When I say that noncognitive attitudes cannot be assessed as justified or unjustified, I mean epistemic justification.

with beliefs concerning actions having moral qualities. Kitcher (2014) calls the view that moral judgments track pre-existing moral properties 'a bad philosophical idea'. He may be correct, and yet exploiting this 'bad idea' might be exactly how ordinary human moral thinking actually functions. After all, how, one might ask, does moral thinking augment social cohesion better than altruistic sentiments? Why is the motivation to cooperate often more reliable when governed by thoughts like 'It is my duty to help him' than when governed by thoughts like 'Gee, I really like him'? The former, it is tempting to answer, gains motivational traction by exploiting the idea (however vaguely) of externally binding rules of conduct — imperatives that are inescapable because they do not depend upon us for their authority — moral truths to which our judgments must conform, not vice versa (Kitcher (2014) refers to the idea of a 'transcendent policeman'). But such ideas will typically accomplish this social role only if they are believed.[8] And so long as they are beliefs, we can immediately ask 'Are they true?' and 'Are they (epistemically) justified?'.

It is the job of the philosopher to investigate whether any sense can be made of this idea of 'inescapable authority'. If it cannot, then human moral beliefs may be systematically false (or, less aggressively: human moral beliefs strictly construed may be systematically false). The fact that one may nevertheless be able to speak of some moral systems serving their evolutionary function better than others — that is, to speak of moral progress — would not cut against this skepticism. To use a crude analogy: Religion may have evolved to serve some social function, and some religions may do so better than others, but for all this atheism may be true.

5. Conclusion

In conclusion let me summarize what this paper has attempted via a quick clarification of two potentially misleading pieces of terminology: 'naturalization' and 'value'.

Most of us seek a naturalization of human morality. We want to understand morality as a non-mysterious phenomenon, with a history that possibly

---

[8] On other occasions I have explored the idea that merely thinking such thoughts, without believing them, might have motivational impact (Joyce, 2001, 2005), but here such complications are bracketed off.

stretches deep into our evolutionary past, accessible to empirical scrutiny. Sections 2 and 3 of this paper sought to contribute (modestly) to this goal by drawing attention to some fairly deep challenges for this program. My intention was not to scupper the project, but to face up honestly to some difficulties confronting it. But there's another kind of 'naturalization' which is a whole new ball game. When metaethicists talk of 'moral naturalism' they typically mean the provision of a naturalization of moral properties. This very different kind of naturalization was the concern of Section 4.

The former kind of naturalization seeks to understand how moral judgment fits into the scientific worldview; the latter kind seeks to understand how moral goodness (etc.). fits into the scientific worldview. Obviously, one can be optimistic about the prospects of the former while highly dubious of the latter. Compare, again, the analogous two ways of understanding what it takes to provide a 'naturalization of religion': one seeks to place religious practices and belief within a scientific worldview; the other would seek to locate God within a scientific worldview.

A matching ambiguity emerges when we talk of 'values'. It is helpful to bear in mind that 'value' is both a verb and a noun. We can investigate what is going on in a human brain when its bearer values something; it is far from obvious that doing so contributes anything to our wondering what things have value (cf. Patricia Churchland's guiding questions: "Where do values come from? How did brains come to care about others?" (2011: 12)). Of course, one might think that the latter is some sort of function of the former (in the same way that the monetary value of things depends on what pecuniary value we are collectively willing to assign them) — but this is a substantive and controversial position in moral philosophy requiring argumentative support. Many metaethicists (and folk!) think, by contrast, that 'value' as a noun is the primary notion, while our valuing activity has the derivative goal of discovering and matching what values exist.

Sections 2 and 3 focused on the origins of moral valuing as an activity: worrying that it will be hard to discern whether the human trait of morally valuing things is an adaptation or a byproduct (Section 2), and concerned that the trait is not, in any case, well-defined (Section 3). Section 4 argued that if moral valuing involves beliefs (as I maintain it does), then it is always reasonable to inquire whether these beliefs are true. To do so is to focus on 'moral value' as a noun — asking whether the facts that are necessary to render moral beliefs true (facts about which actions are forbidden, which are morally good, and so forth) actually obtain. Though on this occasion I have lacked the time to present any arguments, my notes of pessimism have probably been apparent.

References

Alexander, R. (1987). The biology of moral systems. — Aldine de Gruyter, Hawthorne, NY.

Ayala, F. (2010). The difference of being human: morality. — Proc. Natl. Acad. Sci. USA 107: 9015-9022.

Blackburn, S. (1998). Ruling passions. — Oxford University Press, Oxford.

Boehm, C. (2012). Moral origins: the evolution of virtue, altruism, and shame. — Basic Books, New York, NY.

Camus, A. (1951). L'Homme révolté. — Gallimard, Paris.

Churchland, P. (2011). Braintrust: what neuroscience tells us about morality. — Princeton University Press, Princeton, NJ.

Darwin, C. (1879/2004). The descent of man. — Penguin Books, London.

Dennett, D. (1995). Darwin's dangerous idea. — Simon & Schuster, New York, NY.

de Waal, F.B.M. (1992). The chimpanzee's sense of social regularity and its relation to the human sense of justice. — In: The sense of justice: biological foundations of law (Masters, R. & Gruter, M., eds). Sage Publications, Newbury Park, CA, p. 241-255.

de Waal, F.B.M. (1996). Good natured: the origins of right and wrong in primates and other animals. — Harvard University Press, Cambridge, MA.

de Waal, F.B.M. (2006). Primates and philosophers. — Princeton University Press, Princeton, NJ.

Dwyer, S. (2006). How good is the linguistic analogy? — In: The innate mind, Vol. 2: culture and cognition (Carruthers, P., Laurence, S. & Stich, S., eds). Oxford University Press, Oxford, p. 237-255.

Flack, J. & de Waal, F.B.M. (2001). 'Any animal whatever': Darwinian building blocks of morality in monkeys and apes. — In: Evolutionary origins of morality: cross-disciplinary perspectives (Katz, L., ed.). Imprint Academic, Thorverton, p. 1-29.

Garcha, H. & Ettlinger, G. (1979). Object sorting by chimpanzees and monkeys. — Cortex 15: 213-224.

Gould, S.J. & Lewontin, R.C. (1979). The spandrels of San Marco and the Panglossion paradigm: a critique of the adaptationist programme. — Proc. Roy. Soc. Lond. B: Biol. Sci. 205: 581-598.

Greene, J. & Haidt, J. (2002). How (and where) does moral judgment work? — Trends Cogn. Sci. 6: 517-523.

Haidt, J. (2001). The emotional dog and its rational tail: a social intuitionist approach to moral judgment. — Psychol. Rev. 108: 814-834.

Hauser, M., Chomsky, N. & Fitch, W.T. (2002). The faculty of language: what is it, who has it, and how did it evolve? — Science 298: 1569-1579.

Horberg, E., Oveis, C. & Keltner, D. (2011). Emotions as moral amplifiers: an appraisal tendency approach to the influences of distinct emotions upon moral judgment. — Emot. Rev. 3: 237-244.

Irons, W. (1996). Morality as an evolved adaptation. — In: Investigating the biological foundations of human morality (Hurd, J., ed.). Edwin Mellen Press, Lewiston, NY, p. 1-34.

Joyce, R. (2001). The myth of morality. — Cambridge University Press, Cambridge.

Joyce, R. (2005). Moral fictionalism. — In: Fictionalism in metaphysics (Kalderon, M., ed.). Oxford University Press, Oxford, p. 287-313.

Joyce, R. (2006). The evolution of morality. — MIT Press, Cambridge, MA.

Kitcher, P. (2011). The ethical project. — Harvard University Press, Cambridge, MA.

Kitcher, P. (2014). Is a naturalized ethics possible? — Behaviour 151: 245-260.

Krebs, D. (2005). The evolution of morality. — In: The handbook of evolutionary psychology (Buss, D., ed.). Wiley, New York, NY, p. 747-771.

Machery, E. & Mallon, R. (2010). The evolution of morality. — In: The moral psychology handbook (Doris, J., Harman, G., Nichols, S., Prinz, J., Sinnott-Armstrong, W. & Stich, S., eds). Oxford University Press, Oxford, p. 3-46.

Mikhail, J. (2011). Elements of moral cognition: Rawls' linguistic analogy and the cognitive science of moral and legal judgment. — Cambridge University Press, Cambridge.

Nichols, S. (2005). Innateness and moral psychology. — In: The innate mind: structure and contents (Carruthers, P., Laurence, S. & Stich, S., eds). Oxford University Press, New York, NY, p. 353-430.

Prinz, J. (2007). The emotional construction of morals. — Oxford University Press, Oxford.

Prinz, J. (2008). Is morality innate? — In: Moral psychology, Vol. 1: the evolution of morality: adaptations and innateness (Sinnott-Armstrong, W., ed.). MIT Press, Cambridge, MA, p. 367-406.

Prinz, J. (2013). Where do morals come from? A plea for a cultural approach. — In: Empirically informed ethics: morality between facts and norms (Christen, M., Fischer, J., Huppenbauer, M., Tanner, C. & van Schaik, C., eds). Springer, Berlin, Chapter 6.

Small, D. & Lerner, J. (2008). Emotional policy: personal sadness and anger shape judgments about a welfare case. — Politic. Psychol. 29: 149-168.

Valdesolo, P. & DeSteno, D. (2006). Manipulations of emotional context shape moral judgment. — Psychol. Sci. 17: 476-477.

Wheatley, T. & Haidt, J. (2005). Hypnotically induced disgust makes moral judgments more severe. — Psychol. Sci. 16: 780-784.

Section 3: Neuroscience and Development

[When citing this chapter, refer to Behaviour 151 (2014) 281–282]

Introduction

This section explores the interactions between 'genetically' programmed neurobiological substrates and 'epigenetic' developmental mechanisms that constitute the biological basis of moral behavior.

Patricia Churchland discusses the evolution of human morality as rooted in the neurobiology of sociality. Through a comparative analysis of the role played by oxytocin and vasopressin for mammalian caring behavior and how they operate in the brain in association with cortical structures, she proposes that these neurobiological substrates created the platform for sociality and morality. These neural mechanisms are tuned up by learning the social practices of the group. Some of these capabilities may be uniquely developed in the human brain.

Based on neuroscience recent findings, Pierfrancesco Ferrari points out the central role of emotions and empathy in understanding the evolution of human morality. A core element of empathic behavior in human and nonhuman primates is the capacity to internally mimic the behavior of others. During primate evolution, these brain circuits were coopted in the social domain. Thus moral cognition could have emerged as the consequence of emotional processing brain networks, probably involving mirror neuron mechanisms, and of brain regions that, through abstract-inferential processing, evaluate the social context and the value of actions in terms of abstract representations.

Larisa Heiphetz and Liane Young review developmental research on cognitive and social psychology in an attempt to unravel the origin of adults' moral judgment. Adults make moral judgments based on many factors, including harm aversion, and the origins of such judgments lie early in development. Interestingly both children and adults distinguish moral violations from violations of social norms. The authors also discuss the influence of intentions on moral judgment, offering evidence that children and adults alike use information about others' intentions.

Is there a distinction in moral judgments of others dependent on whether they are in-group or out-group individuals? This is an important question addressed by Melanie Killen and Michael Rizzo. Moral judgments require the recognition of intentionality, that is, an attribution of the target's intentions towards another. Most research on the origins of morality has focused on intra-group morality, however. There is new evidence that, beginning early in development, children are able to apply moral concepts to out-group members as well. Research with children provides a window into the complexities of moral judgment and into its evolutionary basis and ontogeny.

The Editors

[When citing this chapter, refer to Behaviour 151 (2014) 283–296]

The neurobiological platform for moral values

Patricia S. Churchland *

Department of Philosophy, University of California, San Diego, La Jolla, CA 92093, USA
*Author's e-mail address: pschurchland@ucsd.edu

Accepted 8 October 2013; published online 27 November 2013

Abstract
What we humans call ethics or morality depends on four interlocking brain processes: (1) caring (supported by the neuroendocrine system, and emerging in the young as a function of parental care); (2) learning local social practices and the ways of others — by positive and negative reinforcement, by imitation, by trial and error, by various kinds of conditioning, and by analogy; (3) recognition of others' psychological states (goals, feelings etc.); (4) problem-solving in a social context. These four broad capacities are not unique to humans, but are probably uniquely developed in human brains by virtue of the expansion of the prefrontal cortex (this formulation is based on Chapter 1 of my book, *Braintrust*: *What neuroscience tells us about morality*).

Keywords
morality, oxytocin, vasopressin, hypothalamus, epigenetics.

1. Where do values come from?[1]

Values are not in the world in the way that seasons or the tides are in the world. This has sometimes provoked the idea that moral values come from the supernatural world. A more appealing hypothesis is that moral values are not other-worldly; rather they are social-worldly. They reflect facts about how we feel and think about certain kinds of social behavior. Those processes are drivers of behavior.

The values of self-survival and self-maintenance are not in the world either. But we are not surprised that they shape the behavior of every animal. No one suggests self-survival values are other-worldly. Instead, it is easy

---

[1] The text that follows is adapted from Chapter 4 in Churchland (2013).

to see how the biological world came to be organized around such values. Unless the genes build a brain that is organized to avoid danger, and seek food and water, the animal will not long survive nor likely reproduce. By contrast, an animal that is wired to care about its own self-maintenance has a better shot at having offspring. So certain self-oriented values are favored by natural selection.

The hallmark of moral values is that they involve self-cost in the care of others. Self-care seems to be in conflict with other-care. How can the neuronal organization to support such values be selected for?

2. The hungry brains of homeotherms

The evolution of the mammalian brain marks the emergence of social values of the kind we associate with morality (this story is probably true of birds too, but for simplicity I shall leaves birds aside for now, regrettably). Sociality appears to have evolved many times, but the flexibility associated with mammalian sociality is strikingly different from the sociality of insects. The evolution of the mammalian brain saw the emergence of a brand new strategy for having babies: the young grow inside the warm, nourishing womb of the female. When mammalian offspring are born, they depend for survival on the mother. So the mammalian brain has to be organized to do something completely new: take care of others in much the way she take cares of herself. So just as I keep myself warm, fed and safe, I keep my babies warm, fed and safe.

Bit by evolutionary bit, over some 70 million years, the self-care system was modified so that care was extended to babies. Now, genes built brains that felt pain when the babies fell out of the nest. Also new, when the babies felt pain owing to cold or separation or hunger, they vocalized. This too caused the mother pain and made her respond to diminish the pain. These new mammalian brains felt pleasure when they were together with their babies, and the babies felt pleasure when they were cuddled up with their mother. They liked being together; they disliked being separated. The pleasure and pain systems were extended to respond to social stimuli.

What was so advantageous about the way early mammal-like reptiles made a living that set the stage for this whole new way of having babies and extending care? The answer is energy sources.

The first reptiles that happened to be *homeotherms* had a terrific advantage — they could hunt at night when the cold-blooded competition was sluggish. Pre-mammals probably feasted on sluggish reptiles lying around waiting for the sun to come up, or at least they could forage without fear of reptilian predators. Homeotherms also managed well in colder climates, thus opening new feeding and breeding ranges.

Homeothermy requires a lot of energy, so warm-blooded animals have to eat about ten times as much as comparably sized poikilotherms (Lane, 2009). If you have to take in a lot of calories to survive, it may help to have a brain that can adapt to new conditions by being smart and flexible. Biologically speaking, it is vastly faster to build brains that can learn prodigiously than to rig a genome that builds brains with reflexes for every contingency that might crop up. To accommodate learning, the genome has to have genes that get expressed to make new protein to add wiring to embody new information. That is much less complex than altering a genome so that it builds a brain that can know at birth how to react in many different circumstances (Quartz & Sejnowski, 1999, 2003). Notice that using a learning strategy to tune up the brain for strategic survival also means that at birth the offspring have only basic reflexes. Mammalian babies are dependent.

Learning requires circuitry that can respond to experience in an adaptive manner yet also work hand in hand with the old motivational, pain and drive systems long in place. Laminar cortex is a remarkable computational solution to the Big Learning problem. It can provide the kind of power and flexibility needed for learning, and also for advantageous planning, and efficient impulse control. Gene duplication allows for the smooth addition of cortical subfields, since the basic recipe for a patch of six-laminar organization of cortex appears to be easily repeatable. Hence size of cortex is expandable in response to ecological pressures.

Exactly how the six-layer cortex emerged from the loosely organized one-to-two layer of reptilian dorsal cortex is largely lost in our ancient past. Nevertheless, comparisons of the brains of different existing species as well as studies of brain development from birth to maturity can tell us a lot (Krubitzer & Kaas, 2005; Krubitzer, 2007). It is known that cortical fields supporting sensory functions vary in size, complexity, and in the connectivity portfolio as a function of a particular mammal's lifestyle and ecological niche. For example, flying squirrels have a very large visual cortical field, whereas the platypus cortex has a tiny visual field but large somatosensory fields. The ghost bat, a nocturnal mammal that relies on precise echo-location to hunt, has a relatively huge auditory field, a small visual field, and a so-matosensory field much smaller than that of the platypus (Krubitzer et al., 2011). Among rodents there are very different styles of moving — flying squirrels, swimming beavers, tree-climbing squirrels, for example. This means that there will also be organizational differences in the parts of the brain that are associated with skilled movement, including motor cortex. In all mammals, frontal cortex is concerned with motor function. In front of the motor regions is prefrontal cortex — areas concerned with control, sociality, and decision-making. All of these cortical fields have rich pathways to and from the whole range of subcortical regions.

Brains are energy hogs, and the calorie intake of homeotherms is high not just to keep body temperature constant, but also to keep their big brains in business. Moreover, because young mammalian are so immature at birth, their calorie intake is especially high. Because mammals eat so much more than reptiles, a given range supports fewer of them. Dozens of lizards can feed quite well on a small patch but a patch that size will support fewer squirrels and even fewer bobcats. The implication for litter size is that the more successful strategy may to produce fewer rather than many offspring, and to invest heavily in their welfare to independence and reproductive maturation.

3. Social bonding

Why do mammalian mothers typically go to great lengths to feed and care for their babies? After all, such care can be demanding, it interferes with feeding, and it can be dangerous. Two central characters in the neurobiological explanation of mammalian other-care are the simple nonapeptides, *oxytocin* and *vasopressin*. The hypothalamus regulates many basic life-functions, including feeding, drinking, and sexual behavior. In mammals, the hypothalamus secretes oxytocin, which triggers a cascade of events with the end result that the mother is powerfully attached to her offspring; she wants to have the offspring close and warm and fed. The hypothalamus also secretes vasopressin, which triggers a different cascade of events so that the mother protects offspring, defending them against predators, for example (Keverne, 2007; Porges & Carter, 2007; Cheng et al., 2010).

The lineage of oxytocin and vasopressin goes back about 500 million-years, long before mammals began to appear. In reptiles these nonapeptides play various roles in fluid regulation and in reproductive processes such as egg-laying, sperm ejection, and spawning stimulation. In mammalian males, oxytocin is still secreted in the testes, and still aids sperm ejaculation. In females it is secreted in the ovaries and plays a role in the release of eggs. In mammals, the roles of oxytocin and vasopressin in both the body and the brain were expanded and modified, along with circuitry changes in the hypothalamus to implement post-natal maternal behavior, including suckling and care (Carter et al., 2008; Young & Alexander, 2012).

During pregnancy, genes in the fetus and in the placenta make hormones that are released into the mother's blood (e.g., progesterone, prolactin, and estrogen). This leads to a sequestering of oxytocin in neurons in the mother's hypothalamus. Just prior to parturition, progesterone levels drop sharply, the density of oxytocin receptors in the hypothalamus increases, and a flood of oxytocin is released from the hypothalamus.

The brain is not the only target of oxytocin, however. It is released also in the body during birth, facilitating the contractions. During lactation, oxytocin is needed for milk ejection, but is also released in the brain of both mother and infant with a calming influence. Assuming the typical background neural circuitry and assuming the typical suite of other resident neurochemicals, oxytocin facilitates attachment of mother to baby. And of baby to mother (Keverne & Curley, 2004; Broad et al., 2006).

Physical pain is a 'protect myself' signal, and these signals lead to corrective behavior organized by self-preservation circuitry. In mammals, the pain system is expanded and modified; protect myself and protect my babies. In addition to a pathway that identifies the kind of pain and locates the site of a painful stimulus, there are pathways responsible for emotional pain, prominently associated with the cingulate cortex, but also subcortical structures such as the amygdala. So when the infant cries in distress, the mother's emotional pain system responds and she takes corrective action. Another cortical area, the insula, monitors the physiological state of the entire body. When you are gently and lovingly stroked, this area sends out 'emotionally-safe' signals (*doing-very-well-now*). The same emotionally-safe signal emerges when the baby is safe and content. And of course the infant responds likewise to gentle and loving touches: *ahhhhh, all is well, I am safe, I am fed.* Safety signals down-regulate vigilance signals such as cortisol. When anxiety and fear are down-regulated, contentment and peacefulness can take their place.

The expression of maternal behavior also depends on the endogenous opiods. This means that during suckling and other kinds of infant care, the opiods down-regulate anxiety, allowing for peaceful responses. If opiod receptors are experimentally blocked, maternal behavior is blocked. This has been observed, for example, in rats, sheep, and rhesus monkeys (Martel et al., 1993; Keverne, 2004; Broad et al., 2006). A reasonable speculation is that the endogenous cannabinoids also play an important role, but much about the extensive cannabinoid system remains unknown.

Although some mammals, such as marmosets and titi monkeys are biparental, in many species, the father takes no interest in parenting and shows none of the mother's attachment to the infant. There are many variations on the basic circuitry regulating parental behavior, depending on a species' ecological niche and how it makes its living. For example, sheep refuse to suckle any lamb that is not their own, whereas pigs and dogs will often suckle nonkin, and even infants of other species.

Studies on rodents of the effect of separation of a pup from the mother (3 h a day for the first two weeks of life) reveal experience-dependent changes in oxytocin and vasopressin synthesis, as well as changes in brain-specific regions of receptors for oxytocin and vasopressin. Behaviorally, the pups that were separated from their mothers showed heightened aggression and anxiety. In some way that is not yet entirely understood, the rats' brains and behavior were altered in a deprived social environment (Veenema, 2012). In a set of important finding on the relationship between stress regulation, genes expression and social behavior, Michael Meaney and colleagues have shown in rodents that during infancy, licking and loving stimulates gene expression that affects the density of receptors for oxytocin in the hypothalamus. More generally, parental tending, or lack thereof, regulates neuroendocrine responses to stress (Meaney, 2001). They also showed that variations in maternal care of female infants is associated with subsequent variations in maternal care displayed by those same females to the next generation of infants. This is a remarkable epigenetic effect. It suggests that neglect or abuse adversely affects the capacity for normal caring, and hence for normal socialization. Further research will explore this matter further.

Here is where we are in the values story: that anything has value *at all* and is motivating *at all* ultimately depends on the very ancient neural organization serving survival and well-being. With the evolution of mammals, the rudimentary 'self-caring organization' is modified to extend the basic values of being alive and well to selected others — to *Me and Mine*. Depending on the evolutionary pressures to which a species is subject, caring may extend to mates, kin, and to friends. Social mammals do tend to show attachment and caring behavior to others besides their own offspring. Exactly which others come within the ambit of caring depends, as always, on the species, how it makes its living, and whether it is thriving. The pain of another's distress and the motivation to care seems to fall off with social distance. By and large, motivation to care seems to be stronger for offspring than for affiliates, for friends than for strangers, for mates than for friends, and so on.

If the maternalization of the brain means that care extends to offspring via mechanisms in the hypothalamus, are those same mechanisms modified to extend care to mates and others? The answer is not entirely clear at this point. Nevertheless, prairie voles (*Microtus ochrogaster*), who tend to bond for life, have provided an important avenue of research on this question. In this context, bonding means that mates prefer the company of each other to that of any other vole. Bonded mates like to be together, the male guards the nest, and they show stress when separated. Male prairie voles also participate in rearing the pups. In prairie voles, permanent bonding typically occurs after the first mating. Bonding does not imply sexual exclusivity, but regardless of other mating interactions, the pair remains as mates that spend a lot of time together and share parenting.

Montane voles, by contrast, do not exhibit comparable social behavior, nor does the male have any role in guarding the nest or rearing the pups. They are not social, and do not like to huddle or hang out with each other.

Because these two species are so very similar, save for their social behavior, the intriguing question is this: what are the relevant differences between the brains of prairie voles and montane voles? It turned out that the differences were not macrostructural. Rather, one major difference is microstructural, pertaining mainly to oxytocin, vasopressin, and differences in the density of receptors that can bind those hormones.

In one region of the reward system (the *nucleus accumbens*), the prairie voles contrast with the montane voles in having a higher density of receptors for oxytocin. In another region of the reward system (*ventral pallidum*) prairie voles have a higher density of receptors for vasopressin. It should also be noted that both males and females have oxytocin and vasopressin, along with their cognate receptors.

The differences in receptor density are one circuit-level difference that help explain long-term attachment of mates after the first mating, but there are other factors involved as well. For example, after mating, the mates need to be able to recognize one another as individuals. Recognition requires learning, which is mediated by the neurotransmitter, dopamine. So if you block the receptors for dopamine, the vole cannot remember whom it was she mated with, and so bonding with a particular mate does not occur. It should also be noted that the receptor density portfolio seen in prairie voles may not extend to all pair-bonders. For example, in mice, the density of vasopressin receptors in the ventral pallidum does not distinguish monogamous from promiscuous species of mice (Goodson, 2013). For technical and ethical reasons, essentially nothing is known about human nonapeptide receptor densities.

Though very common among birds, strong mate preference is somewhat uncommon in mammals. Only about three percent of mammals, including prairie voles, pine voles, California deer mice, beavers, titi monkeys and marmosets show mate attachment.

How exactly do oxytocin and vasopressin regulate other-care? A proper answer would involve the details of all the relevant circuitry and how the neurons in the circuits behave. Unfortunately, these details are not yet known (Goodson, 2013). What is known is that in rodents oxytocin down-regulates the activity of neurons in the amygdala, a structure mediating fear responses and avoidance learning, among other things (Panksepp, 2003; Panksepp & Biven, 2012). When animals are in high alert against danger, when they are preparing to fight or flee, stress hormones are high and oxytocin levels are low. When the threat has passed and the animals is among friends, hugging and chatting, stress hormones back off and oxytocin levels surge. So not only are the amygdala-dependent fear responses down-regulated, but the brain-stem switches from fight-and-flight preparation to rest-and-digest mode.

Is oxytocin the love molecule or the cuddle molecule, as has sometimes been suggested? No. The serious research on oxytocin reveals how very complicated is its action, and how complicated is the circuitry underlying so-cial attachment (Churchland & Winkielman, 2012). Some remarkable claims about correlations between strength of love and blood levels of oxytocin are so astonishing as to raise a flag regarding experimental procedures (McCulloch et al., 2013). Caution is in order.

Lest it be thought that if something is good, more of it will be better, here is a cautionary note. If extra oxytocin is injected into the brain of a happily mated female prairie vole, her degree of mate attachment actually wanes, not rises, and she may become promiscuous.

4. Morality in humans

The foregoing constitutes a very brief overview of what is known about how oxytocin and vasopressin operate in the brain to create a platform for sociality, and hence for morality. But how do we get from a general disposition to care about others, to specific moral actions, such as telling the truth, respecting the goods of others, and keeping promises? How to we get from familial caring to broader community-wide values such as honesty, loyalty and courage? The answer has two intertwined parts: learning by the young, and problem-solving by everyone.

In group-living species such as humans, lemurs and baboons, learning the local conventions and the personality traits of individuals, knowing who is related to whom, and avoiding blackening one's own reputation become increasingly important. Learning, especially by imitation, is the mammalian trick that gets us both flexibility and well-grooved skills. Problem-solving, in the context of learning by trial and error, is the complementary trick that leads to stable social practices for avoiding such problems as conflict.

Children observe, sometimes quite automatically and implicitly, sometimes explicitly and with reflection, the advantages of cooperation. Two children rowing a boat gets them across the lake much faster; two turning the long skipping rope allows doubles skipping, turn-taking means everyone gets a chance so the games do not break down. Men working together can raise a barn in one day. Women working together feed all the men and the children. Singing in a group with parts makes beautiful music. Pitching a tent is easier with two people, and hiking together provides safety. A child quickly comes to recognize the value of cooperation (on the formation of group identity in children, see Killen & Rutland, 2013).

This does not mean that there is a gene 'for cooperation'. If you are sociable, and you want to achieve some goal, then a cooperative tactic can seem a fairly obvious solution to a practical problem. As philosopher David Hume observed, a crucial part of your socialization as a child is that you come to recognize the value of social practices such as cooperation and keeping promises. This means you are then willing to sacrifice something when it is necessary to keep those practices stable in the long run. You may not actually articulate the value of such social practices. Your knowledge of their value may even be largely unconscious, but the value shapes your behavior nonetheless. Brosnan (2011) suggests this is true also of nonhuman primates.

In this context it is important to remember that although all mammals are born immature and learn a great deal during development, the period of human immaturity is especially long and the amount of learning is prodigious. For example, about 50% of a human brain's connections emerge after birth, and the human adult brain weighs about five times that of the infant brain. (Bourgeois, 1997; Huttenlocher & Dabholkar, 1997).

Moreover, in the period leading up to puberty the human brain undergoes substantial pruning and therewith a decrease in connectivity, whereas rodent brains and monkey brains do not show the same degree of pre-pubertal pruning. Jean-Pierre Changeux has argued that these particular epigenetic features of human brain development — extended immaturity and pre-pubertal pruning — enable learning of complex social and cultural organization. (Changeux, 1985). More succinctly, Changeux proposes that the unique developmental profile is what has made human culture, including its moral institutions, possible. Interestingly, this unusually long period of immaturity may depend only on a few regulatory genes that extend the period of epigenetic responsivity to the social and physical environments (Keverne & Curley, 2008).

What I call problem-solving is part of a general capacity to do smart things, and to respond flexibly and productively to new circumstances. Social problem-solving is directed toward finding suitable ways to cope with challenges such as instability, conflict, cheating, catastrophe and resource scarcity. It is probably an extension to the social domain of a broader capacity for problem solving in the physical world. Depending on what you pay most attention to, you may be more skilled in the social domain or in the nonsocial domain, or vice versa. From this perspective, moral problem-solving is, in its turn, a special instance of social problem-solving more broadly (Peterson, 2011).

Although evaluating how to proceed with a particular case is frequently the most pressing concern, the more fundamental problem concerns general principles and institutional structures that undergird well-being and stability.

The development of certain practices as normative — as the right way to handle *this* problem — is critical in a group's cultural evolution (Kitcher, 2012). These norms are established principles enjoining group members against such behavior as embezzlement and other specific forms of cheating. Motivated to belong, and recognizing the benefits of belonging, humans and other highly social animals find ways to get along, despite tension, irritation, and annoyance. Social practices may differ from one group to another, especially when ecological conditions are different. The Inuit of the Arctic will have solved some social problems differently from the Piranhã of the Amazonian basin in Brazil, if only because social problems are not isolated from the physical constraints such as climate and food resources (Hoebel, 1954; Everett, 2009).

Similarities in social practices are not uncommon, as different cultures hit upon similar solutions to particular problems. Subtle and not so subtle differences may also obtain. This is akin to common themes in other practices, such as boat-building or animal husbandry. Particular cultures developed skills for building particular styles of boats — dugout canoes, birch bark canoes, skin-backed kayaks, rafts with sails, junks for fishing on the rivers, and so forth. After many generations, the boats made by separate groups are exquisitely suited to the particular nature of the waters to be traveled on and the materials available. Notice too that many different cultures learned to use the stars for navigation. Some picked up the trick from travelers, others figured it out independently, just as conventions for private property occurred in different groups as their size expanded as agricultural practices became widespread. I am reasonably confident that there is no gene for navigating by the stars.

Though expressions of moral values can vary across cultures, they are not arbitrary, in the way that the conventions for funerals or weddings tend to be. Matters of etiquette, though important for smoothing social interactions, are not serious and momentous as moral values are. Truth-telling and promise-keeping are socially desirable in all cultures, and hence exhibit less dramatic variability than customs at weddings. Is there a gene for these behaviors? Though that hypothesis cannot be ruled out, there is so far no evidence for a truth-telling or a promise-keeping gene. More likely, practices for truth-telling and promise-keeping developed in much the same way as practices for boat building. They reflected the local ecology and are a fairly obvious solution to a common social problem (Hoebel, 1954).

Being reminded of the variability in what counts as morally acceptable helps us acknowledge that standards of morality are not universal. More generally, it reminds us that moral truths and laws do not reside in Plato's heaven to be accessed by pure reason. It reminds us that perorations about morality are often mixed with a whole range of emotions, including fear, resentment, empathy and compassion (Decety, 2011).

5. Tensions, conventions and balance

The mammalian brain is organized both for self-care and to develop care for others, but on many occasions, the two conflict. Social life brings benefits, but it also brings tensions. We compete with siblings and friends for resources and status; we also need to cooperate with them. Some individuals are temperamentally more problematic than others. Sometimes you have to tolerate others who are irritating or noisy or smelly.

Social life can often be very subtle, calling for judgment, not strict adherence to rules. As Aristotle and the Chinese philosopher, Mencius, well realized, you cannot prepare for every contingency or for every situation that may crop up in life. Judgment is essential. Sometimes telling a lie *is* the right thing to do — if it saves the group from a madman threatening to blow up a bomb, for example. Sometimes breaking a promise is the right thing to do — if it prevents a truly terrible catastrophe, such as the meltdown of a nuclear reactor. There are no rules for determining when something is a legitimate exception to prohibitions, such as do not lie, do not break a promise, and do not steal. Children quickly learn about prototypical exceptions, and apply fuzzy-bounded categories rather than hide-bound rules (Killen & Smetana, 2008; Park & Killen, 2010). Balance, as all wise moral philosophers have emphasized, may not be precisely definable, but it is needed to lead a good social and moral life. Not every beggar can be brought home and fed, not all your kidneys can be donated, not every disappointment can be remedied (Schwartz & Sharpe, 2010).

6. Concluding remarks

The capacity for moral behavior is rooted in the neurobiology of sociality, and in mammals depends on nonapeptides oxytocin and vasopressin, as well as on elaborated cortical structures that interface with the more ancient structures mediating motivation, reward, and emotion. The neural mechanisms supporting social behavior are tuned up epigenetically by social interactions and by learning the social practices of the group, and by figuring out how to best deal with new social problems. Emerging after the advent of agriculture and the growth of large groups, organized religions would have built upon existing social practices, perhaps augmenting them in ways relevant to new social demands. Although it is known that oxytocin and vasopressin are critical in social behavior, much about their roles as well as the circuitry with which they interact remains unknown.

References

Bourgeois, J.P. (1997). Synaptogenesis, heterochrony and epigenesis in mammalian neocortex. — Acta Pediatr. Suppl. 422: 27-33.

Broad, K.D., Curley, J.P. & Keverne, E.B. (2006). Mother–infant bonding and the evolution of mammalian social relationships. — Phil. Trans. Roy. Soc. 361: 2199-2214.

Brosnan, S.F. (2011). A hypothesis of the co-evolution between cooperation and response inequity. — Front. Decis. Neurosci. 5: 43.

Carter, C.S., Grippo, A.J., Pournajafi-Nazarloo, H., Ruscio, M. & Porges, S.W. (2008). Oxytocin, vasopressin, and sociality. — Progr. Brain Res. 170: 331-336.

Changeux, J.-P. (1985). Neuronal man. — Pantheon Books, New York, NY.

Cheng, Y., Chen, C., Lin, C.-P., Chou, K.-H. & Decety, J. (2010). Love hurts: an fMRI study. — NeuroImage 51: 923-929.

Churchland, P.S. (2011). Braintrust: what neuroscience tells us about morality. — Princeton University Press, Princeton, NJ.

Churchland, P.S. (2013). Touching a nerve. — Norton, New York, NY.

Churchland, P.S. & Winkielman, P. (2012). Modulating social behavior with oxytocin: How does it work? What does it do? — Horm. Behav. 61: 392-399.

Decety, J. (2011). The neuroevolution of empathy. — Ann. NY Acad. Sci. 1231: 35-45.

Everett, D. (2009). Don't sleep, there are snakes: life and language in the Amazonian jungle. — Pantheon Books, New York, NY.

Goodson, J.L. (2013). Deconstructing sociality, social evolution and relevant nonapeptide functions. — Psychoneuroendocrinology 38: 465-478.

Hoebel, E.A. (1954). The law of primitive man. — Harvard University Press, Cambridge, MA, Chapter 5.

Huttenlocher, P.R. & Dabholkar, A.S. (1997). Regional differences in synaptogenesis in human cerebral cortex. — J. Comp. Neurol. 387: 167-178.

Keverne, E.B. (2004). Understanding well-being in the evolutionary context of brain development. — Phil. Trans. Roy. Soc. Lond. B 359: 1349-1358.

Keverne, E.B. (2007). Genomic imprinting and the evolution of sex differences in mammalian reproductive strategies. — Adv. Genet. 59: 217-243.

Keverne, E.B. & Curley, J.P. (2004). Vasopressin, oxytocin and social behaviour. — Curr. Opin. Neurobiol. 14: 777-783.

Keverne, E.B. & Curley, J.P. (2008). Epigenetics, brain evolution and behavior. — Front. Neuroendocrinol. 29: 398-412.

Killen, M. & Rutland, A. (2013). Children and social exclusion: morality, prejudice and group identity. — Wiley-Blackwell, Oxford.

Killen, M. & Smetana, J.G. (eds) (2006). Handbook of moral development. — Lawrence Erlbaum Associates, Mahwah, NJ.

Kitcher, P.S. (2012). The ethical project. — Harvard University Press, Cambridge, MA.

Krubitzer, L. (2007). The magnificent compromise: cortical field evolution in mammals. — Neuron 2: 201-208.

Krubitzer, L., Campi, K.L. & Cooke, D.F. (2011). All rodents are not the same: a modern synthesis of cortical organization. — Brain Behav. Evol. 78: 51-93.

Krubitzer, L. & Kaas, J. (2005). The evolution of the neocortex in mammals: how is phenotypic diversity generated? — Curr. Opin. Neurobiol. 15: 444-453.

Lane, N. (2009). Hot blood, Chapter 8. — In: Life ascending: the ten great inventions of evolution. W.W. Norton, New York, NY.

Martel, F.L., Nevison, C.M., Rayment, F.D., Simpson, M.J. & Keverne, E.B. (1993). Opioid receptor blockade reduces maternal affect and social grooming in rhesus monkeys. — Psychoneuroendocrinology 18: 307-321.

McCullough, M.E., Churchland, P.S. & Mendez, A.J. (2013). Problems with measuring peripheral oxytocin: Can data on oxytocin and human behavior be trusted? — Neurosci. Behav. Rev. 37: 1485-1492.

Meaney, M.J. (2001). Maternal care, gene expression, and the transmission of individual differences in stress reactivity across generations. — Annu. Rev. Neurosci. 24: 1161-1192.

Panksepp, J. (2003). Feeling the pain of social loss. — Science 302: 237-239.

Panksepp, J. & Biven, L. (2012). The archaeology of mind: neuroevolutionary origins of human emotions. — Norton, New York, NY.

Park, Y. & Killen, M. (2010). When is peer rejection justifiable? Children's understanding across two cultures. — Cogn. Dev. 25: 290-301.

Peterson, D. (2011). The moral lives of animals. — Bloomsbury Press, New York, NY.

Porges, S.W. & Carter, C.S. (2007). Neurobiology and evolution: mechanisms, mediators, and adaptive consequences of caregiving. — In: Self interest and beyond: toward a new understanding of human caregiving (Brown, S.L., Brown, R.M. & Penner, L.A., eds). Oxford University Press, Oxford, p. 53-71.

Quartz, S.R. & Sejnowski, T.J. (1999). The constructivist brain. — Trends Cogn. Sci. 3: 48-57.

Quartz, S.R. & Sejnowski, T.J. (2003). Liars, lovers and heroes. — William Morrow, New York, NY.

Schwartz, B. & Sharpe, K. (2010). Practical wisdom: the right way to do the right thing. — Riverhead Books, New York, NY.

Veenema, A. (2012). Toward understanding how early-life social experiences alter oxytocin- and vasopressin-regulated social behaviors. — Horm. Behav. 61: 304-312.

Young, L. & Alexander, B. (2012). The chemistry between us: love, sex and the science of attraction. — Penguin, New York, NY.

[When citing this chapter, refer to Behaviour 151 (2014) 297–313]

The neuroscience of social relations.
A comparative-based approach to empathy and
to the capacity of evaluating others' action value

Pier F. Ferrari *

Dipartimento di Neuroscienze, Università di Parma, Italy
* Author's e-mail address: pierfrancesco.ferrari@unipr.it

Accepted 15 September 2013; published online 5 December 2013

Abstract

One of the key questions in understanding human morality is how central are emotions in influencing our decisions and in our moral judgments. Theoretical work has proposed that empathy could play an important role in guiding our tendencies to behave altruistically or selfishly. Neurosciences suggest that one of the core elements of empathic behaviour in human and nonhuman primates is the capacity to internally mimic the behaviour of others, through the activation of shared motor representations. Part of the neural circuits involves parietal and premotor cortical regions (mirror system), in conjunction with other areas, such as the insula and the anterior cingulate cortex. Together with this embodied neural mechanism, there is a cognitive route in which individuals can evaluate the social situation without necessary sharing the emotional state of others. For example, several brain areas of the prefrontal cortex track the effects of one's own behaviour and of the value of one's own actions in social contexts. It is here proposed that, moral cognition could emerge as the consequence of the activity of emotional processing brain networks, probably involving mirror mechanisms, and of brain regions that, through abstract-inferential processing, evaluate the social context and the value of actions in terms of abstract representations. A comparative-based approach to the neurobiology of social relations and decision-making may explain how complex mental faculties, such as moral judgments, have their foundations in brain networks endowed with functions related to emotional and abstract-evaluation processing of goods. It is proposed that in primate evolution these brain circuits have been co-opted in the social domain to integrate mechanisms of self-reward, estimation of negative outcomes, with emotional engagement.

Keywords

mirror neurons, neuroeconomics, mimicry, embodiment, orbitofrontal cortex.

1. Introduction

Morality has been for centuries at the centre of philosophical debates mainly because it is considered one of the building blocks of human societies. Indeed, the topic has fuelled strong controversies in philosophy and science since it has profound implications on our understanding of the inner forces that drive and guide behaviours. Within these debates key questions have been addressed in the attempt to understand what characterises our nature from that of other animals, and to which extent we can consider ourselves as unique: is humans' inner nature good or bad? Can human behaviour escape the instincts guiding other animals and selfish genes? Can morality emerge by emancipation from emotions, or are emotions central to our decision processes? Are human altruistic tendencies rational or do they derive from the capacity to share emotions with others? Some philosophers believe that we reason ourselves to moral principles, that emotions are secondary, that morality is all a product of the ratio, whereas others, such as David Hume give moral sentiments (passions and emotions) a central place.

Although it is outside the scopes of this paper to summarise different positions, it is worth noting that reflections on these issues have often suffered from a top-down perspective. In fact very rarely disciplines within the life sciences are given the opportunity to contribute to the debate primarily because of the lack of a proper evolutionary theorisation, and secondly because it was only possible to infer, but not directly examine, internal processes guiding human behaviours.

The idea that human morality cannot escape the rules of evolutionary processes was first proposed by Darwin in the *Descent of Man* (1871): "Besides love and sympathy, animals exhibit other qualities connected with the social instincts, which in us would be called moral.... All animals living in a body, which defend themselves or attack their enemies in concert, must indeed be in some degree faithful to one another; and those that follow the leader must be in some degree be obedient...". According to Darwin, moral decisions are strongly influenced by emotional processes, and in social animals these 'social instincts' are central for the feeling of pleasure when helping others and of unease when harming others.

The work on nonhuman primates has been of great value in challenging the dualistic view of human morality. Several studies have shown that monkeys and apes are capable of reciprocity, are sensitive to others' distress, and can be altruistic without expecting an equal value as return (de Waal, 2008).

The sensitivity to others' emotions indicate that monkeys and apes are capable to empathise with others probably through some basic mechanisms of mirroring or embodied simulation (as described below) which allow individuals to directly access to others' experience (de Waal, 2008; Palagi et al., 2009). The social nature of our species and of our relatives seems to have been inevitably rooted into mechanisms that facilitate the sharing of emotional experiences. The natural tendency to empathise with own group members might translate into behaviours that are indicative of evolved sophisticated altruistic tendencies in highly social species (de Waal, 2012).

Thus, studies on cooperation, consolation, sharing emotions and goods, are shedding light into the inner world of our relatives, and suggest that emotions and empathy not only likely play a major role in the decision-making processes, but are core elements necessary for the development of a moral cognition.

2. Empathy

Empathy is not only the capacity to share and understand others' feeling and emotions, but it is becoming evident that it is a multilayered phenomenon in which emotions and cognitive processes are simultaneously at work (de Waal, 2008; Bernhardt & Singer, 2012). Instead of searching for a unified theory of empathy, several researchers have attempted to dissect it in its core elements and to understand its basic mechanisms in terms of neural underpinnings and cognitive processes.

Several scholars agree that at the basis of empathic responses among several animal species, including humans, there is an emotional response that is shared between two or more individuals, named emotional contagion (Preston & de Waal, 2002; de Vignemont & Singer, 2006). This phenomenon is probably based on an action-perception mechanism and is widespread among primates. Recent work has shown that in humans, apes, and monkeys, yawning is highly contagious (Paukner & Anderson, 2006; Campbell et al., 2009; Palagi et al., 2009; Norscia & Palagi, 2011; Demuru & Palagi, 2012), and its frequency correlates with the quality of the relationship between individuals, suggesting that there is a link between contagious behaviours and interpersonal emotional connection.

These findings also suggest that one of the core elements of empathic behaviour is the capacity to mimic the behaviour of others. This unconscious and automatic phenomenon likely relies on brain mechanisms that facilitate the activation of shared motor representations, which may promote the emergence of a sense of familiarity and emotional connectedness between individuals (Palagi et al., 2009).

Other behavioural phenomena involving body mimicry are common during affiliative social interactions. In a recent study, we demonstrated in gelada baboons (*Theropithecus gelada*) the presence of rapid facial mimicry during play (Mancini et al., 2013) (see Figure 1). Similar behaviours have also been described in apes (Davila-Ross et al., 2007). More interestingly, the speed and frequency of response were higher among individuals with strong bonds, such as mothers and their infants (Mancini et al., 2013). In general, work in humans and nonhuman primates converges in describing a close relationship between emotional contagion, mimicry, and social closeness.

The capacity to mimic others' behaviours and emotions seems to stem from an ancient evolutionary capacity that is already present very early in primate development. For example, human, ape, and monkey neonates are capable of imitating facial gestures displayed by a human model (Meltzoff & Moore, 1977; Ferrari et al., 2006; Bard, 2007). This capacity probably

Figure 1. An example of rapid facial mimicry of full play face in two juvenile gelada baboons (photo by the author). This figure is published in colour in the online version of this journal, which can be accessed via http://booksandjournals.brillonline.com/content/journals/1568539X.

evolved to tune infant's behaviour to that of the mother, thus facilitating the mother-infant relationship and imitative exchanges (Ferrari et al., 2006; Paukner et al., 2012). It is therefore not surprising that, according to several authors, the building blocks of empathy may be found in the early mother-infant relationship (de Waal, 2008; Decety, 2011). The early postnatal period of human and nonhuman primates is, in fact, characterised by intense affective communication between newborns and adults. Such communication involves a constellation of specific behavioural features that include close, sympathetic maternal imitation and elaboration of infant facial communicative and expressive signals (Stern, 1985; Trevarthen & Aitken, 2001), which are unique in anthropoid species. Despite infants' immature brains and limited cognitive skills, they demonstrate active interest in their social world, and show a surprising ability to discriminate adult communicative expressions and to imitate. These complex social exchanges are indicative of the brain's precocious capacity to be tuned with social stimuli and to decode mothers' emotional behaviours through action-perception mechanisms which internally simulate others' emotional states and behaviours (embodied simulation; Gallese, 2001).

Human morality, therefore, can be seen as an expansion of the primate capacity to care for others, which is wired in humans' social brain from its emergence in early ontogeny. The embodiment of interpersonal relations, in particular, has been the privileged route through which our brain can connect with others' brains. These considerations seem to echo in Darwin's observation that "The feeling of pleasure from society is probably an extension of the parental or filial affections, since the social instinct seems to be developed by the young remaining for a long time with their parents" (Darwin, 1871).

3. The neuroscience of empathy: the embodied channel

Although the embodied channel (i.e., the use of shared representations to directly experience and interpret others' behaviour) will be emphasised in the present review, it is important to note that empathy also involves cognitive components. In theorising that empathy is a multilayered phenomenon, several scholars consider the embodied channel central for automatically reproducing others' affective states. However, other layers are built upon and integrated with this core, both from an ontogenetic and evolutionary point of view, which include psychological constructs and cognitive efforts. These latter components are responsible for the capacity of empathising through perspective taking and mentalising, putting oneself in others' shoes.

From a neurobiological standpoint, action-perception mechanisms have been widely investigated in the last few decades and recent advances in neuroscience were particularly stimulated by the discovery of mirror neurons (MN), neurons in the premotor and parietal cortices (see Figure 2) that fire both when a specific action is observed and when the same action is performed by an individual (di Pellegrino et al., 1992; Gallese et al., 1996; Rizzolatti et al., 1996; Ferrari et al., 2003; Fogassi et al., 2005; see Rizzolatti & Craighero, 2004 for a review). The fact that MN have been found in cortical areas involved in motor control, has led to the proposal that others' actions can be translated into a motor code exploiting the inner knowledge, in terms of cortical motor representations of the individual. This translation allows an individual to map others' actions/emotions onto the internal motor representation of that action/emotion. There are several advantages of this action-perception mechanism, compared to other models: (i) it is a parsimonious mechanism to automatically exploit the internal motor knowledge of the individual in order to recognise others' behaviour; (ii) it can explain several phenomena of imitation and of motor social facilitation such as emotional contagion; and (iii) it requires no effort or complex cognitive skills, and is therefore suitable for organisms at an early stage of development, thus supporting several behavioural processes.

The properties of MN make them also a plausible candidate neural mechanism for empathy and social competence: (1) MN fire during observation of facial gestures; (2) the observation of emotional facial gestures also activates areas that are part of the emotional system and that control visceral responses; and (3) the mirror system is involved in imitation.

4. A neural simulation to empathise with others

Although MN were first described as a class of visuomotor neurons discharging for hand actions, more recently, an additional class of MN have been described, which fire for actions performed with the mouth (Ferrari et al., 2003). More interestingly, a small percentage of MN responded while the monkey observed affiliative communicative gestures (i.e., lipsmacking). Similarly, fMRI studies in humans have demonstrated that during the observation and imitation of emotional facial expressions areas of the MN system

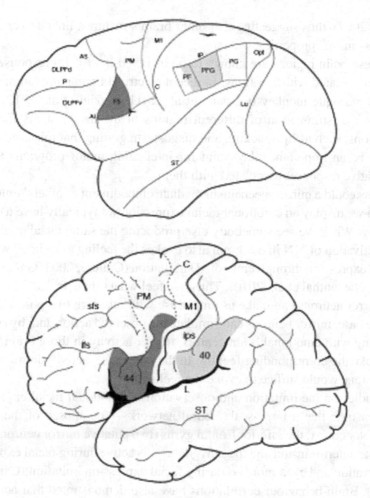

Figure 2. Lateral view of the macaque monkey (top) and human (bottom) cerebral cortex with classical anatomical subdivision showing the areas in which mirror neurons have been found in the monkey and that in the hypothesised homolog areas of humans which: the ventral premotor cortex (red: F5 in the monkey; lower part of the precentral gyrus, area 6, in human), part of the human inferior frontal gyrus (posterior part of area 44), and the inferior parietal lobule (yellow: PFG and part of PF in the monkey and area 40 in humans). This figure is published in colour in the online version of this journal, which can be accessed via http://booksandjournals.brillonline.com/content/journals/1568539X.

are activated (Carr et al., 2003; van der Gaag et al., 2007; Montgomery & Haxby, 2008; Lenzi et al., 2009). In addition to MN areas, the insular cortex and the anterior cingular cortex are activated during the observation of emotional facial expressions (Carr et al., 2003; Wicker et al., 2003; van der Gaag et al., 2007), thus suggesting that other brain structures, linked to emotions, possess mirror properties.

These brain regions are known to be involved in visceral responses typically associated with emotions. In recent electrical brain stimulation studies in the macaque monkey (Caruana et al., 2011; Jezzini et al., 2012) it was shown that stimulation of different regions of the insula elicits facial expressions, such as lipsmacking and disgust, suggesting that this brain region could be an important relay point for integrating motor programs and the vegetative responses associated with them.

How could a mirror mechanism mediate embodiment of others' emotions? When we display an emotional facial expression we typically have a distinct feeling. When we see somebody else producing the same facial expression, the activation of MN in our brain also evokes the feeling associated with that facial expression through embodied simulation (Gallese, 2003; Gallese et al., 2003; Niedenthal et al., 2010). Thus, we feel what others feel.

Mirror neurons can make us empathise with others in two possible ways. In one case, mirror neurons can simulate the observed action and, by communicating with emotional brain centres, trigger activity in those brain centres to evoke the corresponding feeling. In the other case, activity in mirror neurons alone would suffice to evoke the feeling.

Studies on the imitation and observation of emotional facial expressions demonstrate that a large-scale neural network — composed of the ventral premotor cortex, the inferior frontal gyrus (two putative mirror neuron areas), and the anterior insula and the amygdala — is active during facial expression observation and even more so during facial expression imitation (Carr et al., 2003). Brain-behaviour correlations have also demonstrated that activity in this network of areas correlates with measures of empathy and interpersonal skills (Pfeifer et al., 2008). These results seem to suggest that a mirror mechanism could simulate facial expressions, which may evoke simulated activity in limbic structures.

Several studies have investigated the neuroscience of empathy by exploring brain activity while participants witness pain suffered by another person (Singer et al., 2004, 2006; Cheng et al., 2007; Morrison & Downing, 2007; Lamm et al., 2010). In a couple of studies, while undergoing fMRI scans, participants received a painful stimulation or perceived pain in another person (delivered through mild electrical stimulation). In both conditions overlapping regions of the anterior cingulate cortex and the anterior insula were active (Singer et al., 2004, 2006). Similar results were obtained when participants observed images of body parts receiving painful stimulation (Cheng et al., 2007; Lamm et al., 2007; Morrison & Downing, 2007).

Together, these data are compatible with the neural simulation hypothesis, in which an individual can empathise with another person through a process of inner imitation (Iacoboni, 2009). This concept echoes with Theodor Lipps' theoretical account of empathy or Einfühlung (literally it means 'feeling into'), in which the capacity to empathise relies on a mechanism of projecting the self into the other. The psychological process of projection Lipps proposed was based on imitating the inner part of the emotion ('innere Nachahmung' — inner imitation).

The behavioural phenomena ranging from facial mimicry to emotional contagion, and the neurobiological underpinnings described here, reflect automatic responses that do not require complex voluntary control and cognitive efforts. They also demonstrate the close link between the external manifestations of facial/body imitation with the inner imitation of the feeling associated with the emotion.

5. Evaluating goods and others' actions

Work on empathy in humans and nonhuman primates is measuring the degree of sensitivity to others' emotional states. However, work on empathy has also shown that together with the embodied channels there is a cognitive route in which individuals can evaluate the social situation without necessary sharing the emotional state of others. This is especially useful when, for example, individuals make decisions about how to respond in a situation in which their own behaviour could be risky for their own survival (e.g., supporting a companion in a physical conflict with a competitor). In these cases, the empathic mirroring response would be useful for understanding the emotions of others, but the appropriate response should take into account important aspects of the context, with an accurate evaluation of costs and benefits as consequences of one's decision.

Moreover, while engaging in cognitive perspective taking, a person may understand the affective state of another person based on personal previous knowledge, but without sharing the emotion. Accordingly, one can apply to others the same rules that are applied to oneself. This egocentric perspective is very important in social species because, over the course of development, individuals learn how to behave in different social contexts, and what to expect from other individuals when experiencing the same situation. This principle is probably at the basis of important complex behaviours, such as reciprocity and cooperation (de Waal, 2008).

From a neurobiological perspective, we should expect that when making decisions about how to interact with others in social situations (i.e., supporting a conspecific, sharing food) the brain must integrate several pieces of information regarding previous experiences, including the risks and advantages of an action, and it should assign values to others' behaviours or to their own responses (Forbes & Grafman, 2010).

Neuroscientists have only recently explored these issues. In particular, in the last few years neuroscientists have investigated this topic in the monkey brain with clear implications for the field of neuroeconomics as well as for the current theme on morality.

For example, there are areas in the prefrontal cortex that code that value of food received (i.e., juice) and of the physical efforts exerted in order to obtain the food. Some studies have examined the activity of the orbitofrontal cortex (OFC) while the monkey examines different choices requiring an evaluation of food quantity, food type, probability of receiving food, and time delay in the delivery of the food (Padoa-Schioppa, 2011). Surprisingly, different populations of neurons in this area display a discharge that is modulated by all of these factors, suggesting that, in the monkey, the OFC encodes the subjective value of goods, defined in terms of behavioural trade-offs. This same area is also critical in evaluating the rewarding or aversive nature of a stimulus (Morrison & Salzman, 2009), and in comparing the value of different objects.

Similarly, in humans, several brain imaging studies have shown that the OFC and the medial prefrontal cortex (mPFC) are involved in evaluating money amounts, food types, time delays, and probabilities of receiving aversive stimulation (Padoa-Schioppa, 2009; Padoa-Schioppa & Cai, 2011 for a review). Consistent with these findings, lesions to OFC and mPFC result in significant impairments to value-based decision-making (Machado & Bachevalier, 2007; Simmons et al., 2010).

More interestingly, these prefrontal regions seem to maintain an abstract representation of goods that can also be extended to the social domain. For example, in a recent fMRI study, participants made evaluations of moral judgments requiring hypothetical decisions on how to behave in situations in which the sacrifice of one's own life could save the lives of others (Shenhav & Greene, 2010). The ventral medial prefrontal cortex and the ventral striatum were active while participants made these moral judgments.

The role of the prefrontal cortex in evaluating actions within a social domain has also been the target of recent neurophysiological studies in the macaque monkey (Azzi et al., 2012; Chang et al., 2013), in which a monkey performed the task facing other monkeys. In one of these studies (Azzi et al., 2012) each monkey performed the task on alternating days. The task consisted of making an eye gaze fixation and the releasing of a handle, which resulted in receiving a reward. There were two different conditions: in the nonsocial blocks of trials, only the active monkey received the reward, but in the social blocks of trials, both the active and the passive monkeys received the reward. They found that the firing of neurons in the OFC was modulated by the condition, with increased/decreased firing when the reward was social (given to all monkeys), compared to just the active individual. Moreover, the monkeys preferred to deliver joint food with specific partners and were less likely to share rewards when facing more dominant monkeys. OFC neurons track social preferences by enhancing their firing when joint reward is obtained in sessions in which they faced the preferred partner.

Neurophysiological experiments, including those in social situations, show that, in primates, specific brain areas track the effects of one's own behaviour and of the value of one's own actions in social contexts. These cerebral networks probably evolved to encode an abstract representation of goods/values, in value-based decision-making, and have been co-opted in the social domain to mediate the process of self-reward and negative outcomes, motivation, and emotional connections with others.

6. Conclusions

How we can make sense of the neurobiological and behavioural literature on empathy and morality? Neuroscientists have shown that primate brains can empathise with others through a direct body route that involve structures of the MN system, as well as regions that are strictly connected with the visceral response of the organism, while having first-person experiences of others' emotions. Though it is seems likely that this automatic, basic response, shared with several other species, may be the core mechanism supporting empathy, it has also been demonstrated that, built on this core, empathy could have multidimensional layers which require the activity of cognitive perspective–taking mechanisms (Preston & de Waal, 2002; Bzdok et al., 2012; Singer & Hein, 2012). These two systems for empathy have also been demonstrated in patients with brain lesions affecting the inferior frontal gyrus (IFG, part of the MN system) or the ventromedial prefrontal cortex (Shamay-Tsoory et al., 2009). Patients with lesions in the IFG showed impairments in emotional empathy and emotion recognition; while patients with lesions the ventromedial prefrontal cortex (BA 11 and 10) showed deficits in the scores of cognitive empathy.

Though there is a neurological dissociation between these forms of empathy, it is also evident that moral judgments require that the 'emotional/embodied' brain is preserved because it contributes and affects an individual's capacity to represent intentions and actions' values. This is evident by the studies in psychopathic populations, who show impairments in empathic skills, as well as in their capacity to evaluate moral behaviours (Hare, 2003). Interestingly, individuals with autism spectrum disorders also show impairments in embodied empathy (Williams et al., 2006; Baron-Cohen, 2010), deficits in the MN system (Dapretto et al., 2006; Hadjikhani et al., 2007), and difficulties in moral judgments (Zalla et al., 2011; Gleichgerrcht et al., 2012). Conversely, the experience of shared emotions with others in normal individuals, has been shown to reduce aggressive actions towards other people and to promote affiliative behaviours (Eisenberg, 2000).

Together, these empirical accounts suggest that under normal circumstances several autonomous brain networks are simultaneously at work and interacting during decision-making processes. Moral cognition emerges as the consequence of the activity of emotional processing brain networks, probably involving mirror mechanisms, and of brain regions that, through abstract-inferential processing, evaluate the social context and the value of actions in terms of abstract representations.

Several scholars agree on the fact that embodied representations of affect and perspective-taking processes are not mutually exclusive processes, but rather, are integrated (Mitchell, 2005; Keysers & Gazzola, 2007; Singer & Hein, 2012).

From an evolutionary perspective, the neurophysiological work on non-human primates demonstrates that complex mental faculties, such as moral judgments, have their foundations in brain structures endowed with functions related to emotional processing, in the case of emotional contagion, and to the assignment of values to goods through an abstract representation of such goods. Decision-making processes must compute a comparison of different values and guide the selection of a suitable action. These brain areas probably evolved because they were necessary for animals to make decisions about daily activities that were necessary for survival and reproduction, such as the type of food to eat, the amount of time devoted to search, and the cost of an activity. The complexity of social relations in primates (and probably other social species as well) demands a neural architecture that can support necessary cognitive and emotional responses, including empathy. Being cooperative and altruistic require the capacity to understand other emotions and needs, as well as their intentions. It also requires the capacity to evaluate and remember others' actions to reciprocate positive rewarding behaviours, and to avoid potential risky situations. Neurophysiological findings suggest that the same mechanisms involved in economic choice have been recruited for the evaluation of the action value.

A comparative-based approach to the neurobiology of empathy and morality may explain how moral decisions, including those of humans, have emerged through mechanisms shared in the primate lineage, and that the core elements are present both in the emotions and in the rationality. Such an approach is promising for drawing an integrated theory of the nature of human morality.

Acknowledgements

This research was supported by the NIH P01HD064653 grant. I would like to thank Elizabeth Simpson for her comments on an early version of the draft.

References

Azzi, J.C., Sirigu, A. & Duhamel, J.R. (2012). Modulation of value representation by social context in the primate orbitofrontal cortex. — Proc. Natl. Acad. Sci. USA 109: 2126-2131.

Bard, K.A. (2007). Neonatal imitation in chimpanzees (*Pan troglodytes*) tested with two paradigms. — Anim. Cogn. 10: 233-242.

Baron-Cohen, S. (2010). Empathizing, systemizing, and the extreme male brain theory of autism. — Prog. Brain Res. 186: 167-175.

Bernhardt, B.C. & Singer, T. (2012). The neural basis of empathy. — Annu. Rev. Neurosci. 35: 1-23.

Bzdok, D., Schilbach, L., Vogeley, K., Schneider, K., Laird, A.R., Langner, R. & Eickhoff, S.B. (2012). Parsing the neural correlates of moral cognition: ALE meta-analysis on morality, theory of mind, and empathy. — Brain Struct. Funct. 217: 783-796.

Campbell, M.W., Carter, J.D., Proctor, D., Eisenberg, M.L. & de Waal, F.B.M. (2009). Computer animations stimulate contagious yawning in chimpanzees. — Proc. Roy. Soc. Lond. B: Biol. Sci. 276: 4255-4259.

Carr, L., Iacoboni, M., Dubeau, M.C., Mazziotta, J.C. & Lenzi, G.L. (2003). Neural mechanisms of empathy in humans: a relay from neural systems for imitation to limbic areas. — Proc. Natl. Acad. Sci. USA 100: 5497-5502.

Caruana, F., Jezzini, A., Sbriscia-Fioretti, B., Rizzolatti, G. & Gallese, V. (2011). Emotional and social behaviors elicited by electrical stimulation of the insula in the macaque monkey. — Curr. Biol. 21: 195-199.

Chang, S.W., Gariépy, J.F. & Platt, M.L. (2013). Neuronal reference frames for social decisions in primate frontal cortex. — Nature Neurosci. 16: 243-250.

Cheng, Y., Lin, C.P., Liu, H.L., Hsu, Y.Y., Lim, K.E., Hung, D. & Decety, J. (2007). Expertise modulates the perception of pain in others. — Curr. Biol. 17: 1708-1713.

Dapretto, M., Davies, M.S., Pfeifer, J.H., Scott, A.A., Sigman, M., Bookheimer, S.Y. & Iacoboni, M. (2006). Understanding emotions in others: mirror neuron dysfunction in children with autism spectrum disorders. — Nature Neurosci. 9: 28-30.

Darwin, C. (1871). The descent of man. — Murray, London.

Davila Ross, M., Menzler, S. & Zimmermann, E. (2007). Rapid facial mimicry in orangutan play. — Biol. Lett. 4: 27-30.

Decety, J. (2011). The neuroevolution of empathy. — Ann. NY Acad. Sci. 1231: 35-45.

Demuru, E. & Palagi, E. (2011). In bonobos yawn contagion is higher among kin and friends. — PLoS ONE 7: e49613, doi:10.1371/journal.pone.0049613.

de Vignemont, F. & Singer, T. (2006). The empathic brain: how, when and why? — Trends Cogn. Sci. 10: 435-441.

de Waal, F.B.M. (2008). Putting the altruism back into altruism: the evolution of empathy. — Annu. Rev. Psychol. 59: 279-300.

de Waal, F.B.M. (2012). The antiquity of empathy. — Science 336: 874-876.

di Pellegrino, G., Fadiga, L., Fogassi, L., Gallese, V. & Rizzolatti, G. (1992). Understanding motor events: a neurophysiological study. — Exp. Brain Res. 91: 176-180.

Eisenberg, N. (2000). Emotion, regulation, and moral development. — Annu. Rev. Psychol. 51: 665-697.

Ferrari, P.F., Gallese, V., Rizzolatti, G. & Fogassi, L. (2003). Mirror neurons responding to the observation of ingestive and communicative mouth actions in the monkey ventral premotor cortex. — Eur. J. Neurosci. 17: 1703-1714.

Ferrari, P.F., Visalberghi, E., Paukner, A., Fogassi, L., Ruggiero, A. & Suomi, S.J. (2006). Neonatal imitation in rhesus macaques. — PLoS Biol. 4: 1501-1508.

Fogassi, L., Ferrari, P.F., Gesierich, B., Rozzi, S., Chersi, F. & Rizzolatti, G. (2005). Parietal lobe: from action organization to intention understanding. — Science 308: 662-667.

Forbes, C.E. & Grafman, J. (2010). The role of the human prefrontal cortex in social cognition and moral jusdgement. — Annu. Rev. Neurosci. 33: 229-324.

Gallese, V. (2001). The "shared manifold" hypothesis: from mirror neurons to empathy. — J. Conscious. Stud. 8: 33-50.

Gallese, V. (2003). The manifold nature of interpersonal relations: the quest for a common mechanism. — Phil. Trans. Royal Soc. London B 358: 517-528.

Gallese, V., Fadiga, L., Fogassi, L. & Rizzolatti, G. (1996). Action recognition in the premotor cortex. — Brain 119: 593-609.

Gallese, V., Ferrari, P.F. & Umiltà, M.A. (2003). The mirro matching system: a shared manifold of intersubjectivity. — Behav. Brain Sci. 25: 35-36.

Hadjikhani, N., Joseph, R.M., Snyder, J. & Tager-Flusberg, H. (2007). Abnormal activation of the social brain during face perception in autism. — Hum. Brain Mapp. 28: 441-449.

Hare, R.D. (2003). The hare psychopathy checklist-revised. — Multi-Health Systems, Toronto, ON.

Iacoboni, M. (2009). Imitation, empathy, and mirror neurons. — Annu. Rev. Psychol. 60: 653-670.

Jezzini, A., Caruana, F., Stoianov, I., Gallese, V. & Rizzolatti, G. (2012). Functional organization of the insula and inner perisylvian regions. — Proc. Natl. Acad. Sci. USA 109: 10077-10082.

Keysers, C. & Gazzola, V. (2007). Integrating simulation and theory of mind: from self to social cognition. — Trends Cogn. Sci. 11: 194-196.

Lamm, C., Decety, J. & Singer, T. (2010). Meta-analytic evidence for common and distinct neural networks associated with directly experienced pain and empathy for pain. — Neuroimage 54: 2492-2502.

Lamm, C., Nusbaum, H.C., Meltzoff, A.N. & Decety, J. (2007). What are you feeling? Using functional magnetic resonance imaging to assess the modulation of sensory and affective responses during empathy for pain. — PLoS ONE 2: e1292.

Lenzi, D., Trentini, C., Pantano, P., Macaluso, E., Lenzi, G.L. & Ammaniti, M. (2009). Attachment models affect brain responses in areas related to emotions and empathy in nulliparous women. — Hum. Brain Mapp. 34: 1399-1414.

Machado, C.J. & Bachevalier, J. (2007). The effects of selective amygdala, orbital frontal cortex or hippocampal formation lesions on reward assessment in nonhuman primates. — Eur. J. Neurosci. 25: 2885-2904.

Mancini, G., Ferrari, P.F. & Palagi, E. (2013). Rapid facial mimicry in geladas. — Sci. Rep. 3: 1527, doi:10.1038/srep01527.

Meltzoff, A.N. & Moore, M.K. (1977). Imitation of facial and manual gestures by human neonates. — Science 198: 75-78.

Mitchell, J.P. (2005). The false dichotomy between simulation and theory-theory: the argument's error. — Trends Cogn. Sci. 9: 363-364.

Montgomery, K.J. & Haxby, J.V. (2008). Mirror neuron system differentially activated by facial expressions and social hand gestures: a functional magnetic resonance imaging study. — J. Cogn. Neurosci. 20: 1866-1877.

Morrison, I. & Downing, P.E. (2007). Organization of felt and seen pain responses in anterior cingulate cortex. — Neuroimage 37: 642-651.

Morrison, S.E. & Salzman, C.D. (2009). The convergence of information about rewarding and aversive stimuli in single neurons. — J. Neurosci. 29: 11471-11483.

Niedenthal, P.M., Mermillod, M., Maringer, M. & Hess, U. (2010). The Simulation of Smiles (SIMS) model: embodied simulation and the meaning of facial expression. — Behav. Brain Sci. 33: 417-433.

Norscia, I. & Palagi, E. (2011). Yawn contagion and empathy in *Homo sapiens*. — PLoS ONE 6: e28472.

Padoa-Schioppa, C. (2009). Range-adapting representation of economic value in the orbitofrontal cortex. — J. Neurosci. 29: 14004-14014.

Padoa-Schioppa, C. (2011). Neurobiology of economic choice: a good-based model. — Annu. Rev. Neurosci. 32: 333-359.

Padoa-Schioppa, C. & Cai, X. (2011). The orbitofrontal cortex and the computation of subjective value: consolidated concepts and new perspectives. — Ann. NY Acad. Sci. 1239: 130-137.

Palagi, E., Leone, A., Mancini, G. & Ferrari, P.F. (2009). Contagious yawning in gelada baboons as a possible expression of empathy. — Proc. Natl. Acad. Sci. USA 106: 19262-19267.

Paukner, A. & Anderson, J.R. (2006). Video-induced yawning in stumptail macaques (*Macaca arctoides*). — Biol. Lett. 2: 36-38.

Paukner, A., Ferrari, P.F. & Suomi, S.J. (2012). A comparison of neonatal imitation abilities in human and macaque infants. — In: Navigating the social world: a developmental perspective (Banaji, M.R. & Gelman, S.A., eds). Oxford University Press, Oxford, p. 133-138.

Pfeifer, J.H., Iacoboni, M., Mazziotta, J.C. & Dapretto, M. (2008). Mirroring others' emotions relates to empathy and interpersonal competence in children. — Neuroimage 39: 2076-2085.

Preston, S.D. & de Waal, F.B.M. (2002). Empathy: its ultimate and proximate bases. — Behav. Brain Sci. 25: 1-20.

Rizzolatti, G. & Craighero, L. (2004). The mirror-neuron system. — Annu. Rev. Neurosci. 27: 169-192.

Rizzolatti, G., Fadiga, L., Gallese, V. & Fogassi, L. (1996). Premotor cortex and the recognition of motor actions. — Brain Res. Cogn. Brain Res. 3: 131-141.

Shamay-Tsoory, S.G., Aharon-Peretz, J. & Perry, D. (2009). Two systems for empathy: a double dissociation between emotional and cognitive empathy in inferior frontal gyrus versus ventromedial prefrontal lesions. — Brain 132: 617-627.

Shenhav, A. & Greene, J.D. (2010). Moral judgments recruit domain-general valuation mechanisms to integrate representations of probability and magnitude. — Neuron 67: 667-677.

Simmons, J.M., Minamimoto, T., Murray, E.A. & Richmond, B.J. (2010). Selective ablations reveal that orbital and lateral prefrontal cortex play different roles in estimating predicted reward value. — J. Neurosci. 30: 15878-15887.

Singer, T. & Hein, G. (2012). Human empathy through the lens of psychology and social neuroscience. — In: The primate mind (de Waal, F.B.M. & Ferrari, P.F., eds). Harvard University Press, Cambridge, p. 158-174.

Singer, T., Seymour, B., O'Doherty, J., Kaube, H., Dolan, R.J. & Frith, C.D. (2004). Empathy for pain involves the affective but not sensory components of pain. — Science 303: 1157-1162.

Singer, T., Seymour, B., O'Doherty, J.P., Stephan, K.E., Dolan, R.J. & Frith, C.D. (2006). Empathic neural responses are modulated by the perceived fairness of others. — Nature. 439: 466-469.

Stern, D. (1985). The interpersonal world of the infant: a view from psychoanalysis and developmental psychology. — Basic Books, New York, NY.

Trevarthen, C. & Aitken, K.J. (2001). Infant intersubjectivity: research, theory, and clinical applications. — J. Child Psychol. Psychiatry 42: 3-48.

van der Gaag, C., Minderaa, R.B. & Keysers, C. (2007). Facial expressions: what the mirror neuron system can and cannot tell us. — Soc. Neurosci. 2: 179-222.

Wicker, B., Keysers, C., Plailly, J., Royet, J.P., Gallese, V. & Rizzolatti, G. (2003). Both of us disgusted in My insula: the common neural basis of seeing and feeling disgust. — Neuron 40: 655-664.

Williams, J.H., Waiter, G.D., Gilchrist, A., Perrett, D.I., Murray, A.D. & Whiten, A. (2006). Neural mechanisms of imitation and 'mirror neuron' functioning in autistic spectrum disorder. — Neuropsychologia 44: 610-621.

Zalla, T., Barlassina, L., Buon, M. & Leboyer, M. (2011). Moral judgment in adults with autism spectrum disorders. — Cognition 121: 115-126.

[When citing this chapter, refer to Behaviour 151 (2014) 315–335]

Review

A social cognitive developmental perspective on moral judgment

Larisa Heiphetz * and **Liane Young**

Department of Psychology, Boston College, 140 Commonwealth Avenue, Chestnut Hill, MA 02467, USA
*Corresponding author's e-mail address: larisa.heiphetz@bc.edu

Accepted 16 August 2013; published online 5 December 2013

Abstract

Moral judgment constitutes an important aspect of adults' social interactions. How do adults' moral judgments develop? We discuss work from cognitive and social psychology on adults' moral judgment, and we review developmental research to illuminate its origins. Work in these fields shows that adults make nuanced moral judgments based on a number of factors, including harm aversion, and that the origins of such judgments lie early in development. We begin by reviewing evidence showing that distress signals can cue moral judgments but are not necessary for moral judgment to occur. Next, we discuss findings demonstrating that both children and adults distinguish moral violations from violations of social norms, and we highlight the influence of both moral rules and social norms on moral judgment. We also discuss the influence of actors' intentions on moral judgment. Finally, we offer some closing thoughts on potential similarities between moral cognition and reasoning about other ideologies.

Keywords

harm aversion, moral cognition, social cognitive development.

1. A social cognitive developmental perspective on moral judgment

Moral judgment — reasoning about whether our own and others' actions are right or wrong — is a fundamental aspect of human cognition, informing a variety of social decisions. This paper investigates the origins of adults' moral judgments, focusing on moral judgments in the domain of harm. We begin by discussing distress signals that could indicate that harm has occurred and could therefore serve as strong elicitors of moral judgment. We

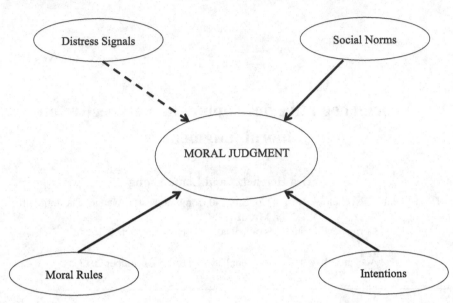

Figure 1. Four inputs to moral judgment. Note that though distress signals may influence moral judgment, such judgment can occur even in the absence of distress signals.

challenge the notion that such signals are required for moral judgments by discussing research demonstrating that moral judgment often occurs in their absence. We then turn to a discussion of 'social domain theory' and present evidence showing that children, like adults, distinguish moral violations from violations of social norms and that social norms can influence moral judgment. Next, we discuss research on 'theory of mind' showing that the moral judgments of neurotypical children and adults depend on information about others' intentions (Figure 1). Finally, we discuss links between moral cognition and other domains.

2. The role of others' distress in moral judgment

Adults typically experience emotional aversion when asked to perform harmful actions such as discharging a gun in someone's face (Cushman et al., 2012). Harm aversion is so common that it may appear, at first glance, to constitute the entirety of moral cognition. Though Graham and colleagues (2011) have identified additional moral domains, harm appears important across a broad spectrum of participants. Unlike other domains that vary in importance across demographic categories, harm influences cognition across diverse cultures (Haidt, 2012), among liberals and conservatives (Graham et al., 2009), and even among some non-human primates (Sheskin & Santos, 2012). Indeed, some definitions of morality include only the domain of harm. For example, de Waal (this issue) defines morality as "helping or at least not hurting fellow human beings".

For this review, we define harms as acts that injure others physically, emotionally, and/or materially. People might reason about different kinds of harm in different ways, yet a variety of actions (e.g., hitting, name-calling, stealing) can still be considered harmful. We use the term "harm aversion" to refer to moral condemnation of harmful actions. We focus on moral judgment rather than moral behavior because the bulk of current research in moral psychology focuses on the former. We argue that harm aversion plays a large role in moral judgment across development but that moral judgment depends on other factors as well. Specifically, both children and adults consider additional aspects of the situation, such as relevant social norms and information about an actor's intentions. Although we focus on the developmental origins of moral judgment, much research has also investigated morality's evolutionary origins (see Boehm, this issue; de Waal, this issue; Joyce, this issue).

One of the clearest ways to tell that harm has occurred is by observing victims' expressions of pain. Researchers have argued that people have evolved to respond to distress signals by ceasing aggression. For example, building on the work of ethologists such as Eibl-Eibesfeldt (1970) and Lorenz (1966), psychologist James Blair proposed that people who are healthy have a violence inhibition mechanism (VIM) that is activated by cues such as crying (Blair, 1995; Blair & Morton, 1995). In other words, crying signals that harm is occurring and should be stopped. Blair argues that the VIM has a long evolutionary history; for example, dogs typically do not kill opponents who bare their throats in a fight, suggesting that some mechanisms have evolved to prevent death in the midst of conflict. Similarly, most people inhibit aggression in the face of others' distress.

Investigations of infants have found evidence consistent with the idea that harm — and signals that harm has occurred — is aversive early in development. For instance, newborns cry in response to another infant's cries (Sagi & Hoffman, 1976; Martin & Clark, 1982; see Ainsworth et al., 1978, for a discussion of the functions of infant crying). Suggestive as it is, such evidence is open to multiple interpretations. For instance, infants may experience empathy at another's distress (regardless of whether or not they find the distress itself aversive), or they may simply be irritated at a disturbance of the peace, or they may infer the presence of a threat in their own environment. Because infants are limited in the types of responses they can provide, it is difficult to disentangle these possibilities among this population. However, evidence from older toddlers can help identify the role these potential factors may play later in development.

One way to obtain such evidence is to observe older infants' and toddlers' provisions of comfort to those who have experienced harm. Such behavior is not specific to humans (Romero et al., 2010; Romero & de Waal, 2010), showing the long evolutionary history of comforting others. Work with human children has shown that neither 18-month-old nor 24-month-old infants comforted an adult who expressed physical pain after hurting her knee, though a minority of participants in both age groups exhibited self-soothing behaviors such as sucking their thumb or touching their own knee (Dunfield et al., 2011). Slightly older toddlers did show greater concern for victims of harm (Zahn-Waxler et al., 1992). Although some two-year-olds may be predisposed to respond to others' distress with empathy (Nichols et al., 2009), toddlers of this age do not always seek to redress those harms in ways typically used by older children and adults (e.g., comforting). It is not until the age of three that toddlers reliably comfort adults who have been harmed by others by sharing their own resources or making suggestions for how the adult can feel better. Toddlers of this age also attempt to prevent harms by telling agents performing negative behaviors to stop (Vaish et al., 2011).

At this point, it may be useful to consider what factors constitute the signatures or precursors of moral judgment for young children. How might we know when children make moral judgments? Self-soothing does not appear to be a strong cue, as this behavior can indicate self-oriented goals that do not concern morality. Comforting others may serve as a better cue in later toddlerhood, but these behaviors do not tend to emerge until around age three and may also indicate self-oriented goals. For example, older toddlers may comfort others to put an end to the distress signals that they find aversive.

Attempting to stop harmful behaviors, as three-year-old toddlers in Vaish and colleagues' (2011) work did, may serve as a stronger indication that toddlers are making a moral judgment. Such behavior (i.e., confronting the harm-doer) may lead to distress in the perpetrator of the harm and is therefore unlikely to serve a self-oriented goal of ending all distress cues. Rather, three-year-old toddlers who attempt to stop harmful behaviors may be indicating that they find these behaviors morally objectionable. Yet other explanations are possible in this case as well, since even non-human animals can attempt to stop behaviors that harm others. For example, bears can aggress against those harming their cubs, and dogs can attempt to prevent harm from occurring to their owners. Such interventions can occur for a number of reasons, including kin selection and human training (e.g., dogs may be trained to attack people or animals who attempt to harm their owners), and may not always indicate moral judgment. Perhaps due to the difficulty of inferring moral judgment from behavior, much work in moral psychology has relied on asking direct questions (e.g., whether a particular behavior was okay or not okay).

3. Evidence that distress signals are not necessary for moral judgment

The previous section describes instances in which young toddlers comfort the victim or confront the harm-doer, but recent work reveals instances when young toddlers and even infants perceive and respond to immoral actions in the absence of distress signals (see Hamlin, 2012, for a review). Thus, early moral cognition may be more nuanced than a simple formula in which "distress signals = harm = immorality". Indeed, infants and young toddlers may have some understanding of morality despite their failure to exhibit this understanding through their behaviors. Individuals in these age groups may understand some behaviors to be immoral but fail to act on this understanding in ways typical among older individuals (e.g., comforting victims).

In one series of studies, Hamlin and colleagues (Hamlin et al., 2007, 2010; Hamlin & Wynn, 2011) showed infants displays featuring "helpers" (e.g., one character facilitating another character's goal to climb up a hill) and "hinderers" (e.g., one character preventing another character from climbing up a hill). As indicated by a variety of measures, including reaching and looking time, infants preferred helpers to hinderers. This occurred even though the characters were portrayed by shapes lacking emotional expressions (e.g., distress). In a different study, two-year-old toddlers showed greater concern for an adult whose property had been destroyed or taken away even if the adult did not display any emotional distress (Vaish et al., 2009). These studies show that infants and toddlers are sensitive to harm even when victims have not indicated they have experienced harm or are in distress.

In summary, even infants appear to distinguish help from harm. Infants and toddlers do not require visible signs of distress to infer that harm has occurred; rather, they prefer helpers over hinderers even in the absence of distress signals. Though infants are unable to articulate their internal states, their preferences for helpers have been interpreted as a form of moral judgment (e.g., Hamlin, 2012). Below we discuss evidence that social norms and moral rules can also impact moral judgment in the absence of distress signals, providing further evidence that such signals are not necessary for moral judgment to occur.

4. The role of norms and rules in moral judgment

4.1. The role of social norms in moral judgment

Much of human behavior, like the behavior of some non-human animals (e.g., dogs: Bekoff, 2001; and monkeys: de Waal, 1993), is influenced by social norms. Toddlers seem to acquire an understanding of norms around three years of age. At this milestone, they begin to infer that actions are normative (that is, they "should be" done a particular way) when an adult simply demonstrates the action with familiarity, even in the absence of pedagogical or language cues (Schmidt et al., 2011). Furthermore, three-year-old toddlers protest when actors violate context-dependent norms such as the rules of a particular game (Rakoczy, 2008; Rakoczy et al., 2008, 2009; Wyman et al., 2009). Three-year-old toddlers accept norms concerning sharing, though they fail to follow these norms themselves. For example, they report that they and others should share stickers equally, though children do not typically distribute valued resources equally until they reach approximately 7 years of age (Smith et al., 2013). Three- and four-year-olds also tattle to authority figures when their siblings and classroom peers violate rules and norms (Ross & den Bak-Lammers, 1998); in one line of work, such tattling represented the majority of children's statements about their peers to third parties (Ingram & Bering, 2010). Toddlers who tattled on siblings tended to emphasize harmful actions such as physical aggression (den Bak & Ross, 1996), suggesting that even toddlers may view rules against harm as especially important, or at least recognize that their parents may take this view.

In many instances, toddlers distinguish social norms from moral rules, which are proscriptions against behaviors that result in negative outcomes towards others (Lockhart et al., 1977; Nucci, 1981; Smetana, 1981; Turiel,

1983; Smetana et al., 1993; for a more thorough review of children's differentiation of social norms from moral rules, see Killen & Rizzo, this issue). For example, three-year-old toddlers enforce moral rules equally for in- and out-group members but enforce conventional norms more for in-group members (Schmidt et al., 2012). In Schmidt and colleagues' study, toddlers met puppets speaking in a native or foreign accent and then saw these puppets violate either moral rules (e.g., damage someone's property) or conventional norms (e.g., play a game the "wrong" way). Toddlers protested equally when actors violated moral rules regardless of group membership but protested more when in-group rather than out-group actors violated conventional norms, demonstrating an understanding that group membership likely exerts a stronger influence on conventional norms.

Although toddlers distinguish conventional social norms from moral rules, they also use information about the former to inform their evaluations of the latter. For example, in one study (Hepach et al., 2013), three-year-old toddlers played a drawing game with two experimenters. One experimenter showed the second experimenter how to cut a piece of paper in one of three ways: (1) cutting a blank piece of paper, (2) cutting a small section of the second experimenter's paper without destroying the drawing on the paper made by the second experimenter, or (3) cutting across the entire drawing made by the second experimenter. The second experimenter then displayed emotional distress. Three-year-old toddlers displayed concern for the second experimenter only when she appeared justifiably upset, i.e., when her picture had been destroyed in the third condition. Specifically, children checked up on the experimenter and helped her with a subsequent task. Children responded with similar levels of concern when they were not privy to information about why the experimenter was upset; that is, without specific evidence of unjustified distress, children assumed the response to be justified. However, when the experimenter responded with strong distress to a minor harm (e.g., when the paper was cut, but the drawing was left intact), children showed significantly less sympathy. That is, in a situation where "victims" expressed great distress in response to a socially normative action (e.g., cutting a small piece of paper), toddlers appeared to view the distress as unjustified. Preschoolers with autism showed this same "crybaby effect", by sympathizing more in the case of justified distress; this aspect of moral cognition may thus be spared despite other social-cognitive impairments among individuals with autism (Leslie et al., 2006).

4.2. The role of rules in moral judgment

In addition to social norms, which are perceived to vary across contexts (Smetana, 1981; Turiel, 1983; Smetana et al., 1993), other more broadly applicable rules can also govern moral judgment. In fact, both Piaget (1932) and Kohlberg (1969) proposed that during the early stages of moral development, moral judgment is primarily rule-based. For example, a child whose moral reasoning is at the pre-conventional stage (in Kohlberg's terminology) might claim that it is wrong for a man to steal a drug that will save his wife's life because stealing is against the rules.

More recent work has investigated the importance of rules to older children and adults. In one study (Lahat et al., 2012), for example, 8–10-year-old children, 12–13-year-old adolescents, and undergraduates read identical stories with endings slightly altered to create violations of moral rules or violations of social norms. For example, in one story, Alice saw her sister's pajamas in her closet. Alice then decided to shred them (moral violation) or wear them to school (conventional violation). Participants were asked to press one key as quickly as possible if they thought the behavior was "OK" and a different key if they thought the behavior was "NOT OK". During half of the trials, participants were instructed to imagine that there was no rule against the behavior (rule removed condition). During the other half, participants were given no specific instructions and were assumed to operate as though a rule prohibited the violations (rule assumed condition). In the rule assumed condition, participants of all ages judged moral violations to be wrong more quickly than they judged conventional violations to be wrong; in the rule removed condition, participants responded equally quickly across these conditions. This suggests that judgments concerning conventional violations require additional cognitive processing and that the presence of rules can alter the ease with which people make moral judgments.

Additionally, adults responded more quickly that moral violations were wrong in the rule assumed condition. Lahat et al. (2012) offer two explanations for this finding. First, adults may have been surprised by the lack of rules against moral violations. Second, adults may have considered the context in which moral violations took place. This latter explanation is at odds with social domain theory (Smetana, 2006), which argues that context influences only judgments concerning conventional violations, not judgments concerning moral violations. However, this interpretation is in line with additional research drawing on the philosophy literature.

Specifically, Nichols & Mallon (2006) drew on the famous trolley problem to investigate the role of rules in moral judgment. Two versions of this dilemma exist. In the 'bystander' case, individuals are asked to imagine a person standing by a trolley track. This bystander sees five people working on the tracks and also notices a train heading directly toward them. If the bystander does nothing, the train will kill the people on the track. However, the bystander has the option to flip a switch, causing the train to switch to another track and kill a sole individual there. Participants are typically asked whether it is morally acceptable to flip the switch. The 'footbridge' case presents a similar dilemma, with one twist: now there is no switch to pull. Rather, the individual observing the train has the option of pushing another individual (typically a large stranger) in front of the train. This action would kill the stranger but save the lives of the people working on the track. In both cases, claiming that it is acceptable to sacrifice one life to save five reflects 'utilitarianism'. Although the utilitarian option ultimately results in more saved lives, several philosophers have asserted that it is not always the moral option (Thomson, 1976; Quinn, 1989), and this non-utilitarian intuition has been strengthened by presenting the dilemma in terms of the trolley problem rather than in abstract terms.

In recent empirical work, adults made different judgments in the bystander scenario and the footbridge scenario. Healthy adults are more likely to endorse the utilitarian option in the bystander scenario than in the footbridge scenario (Greene et al., 2001; Shallow et al., 2011; Côté et al., 2013). This difference may reflect an emotion-based aversion to harming others via direct physical contact (Greene et al., 2001). Most people hold an intuition that harming others is wrong, and they may search for cognitive reasons to justify this emotional intuition when presented with the footbridge scenario (Hume, 1739; Haidt, 2001, 2012).

To investigate the role of context on moral judgments, Nichols & Mallon (2006) altered the basic bystander and footbridge scenarios by presenting participants with stories featuring "victims" that were teacups rather than people. These scenarios included the same cost: benefit ratios as traditional trolley dilemmas (e.g., sacrificing five to save one); however, these ratios are applied to inanimate objects. In both scenarios, a mother tells her child not to break any teacups and then leaves. A situation then occurs where a toy vehicle is likely to run over multiple teacups. The child saves multiple teacups either by diverting the train away from multiple cups and causing it to crush one solitary cup ("bystander" scenario) or by throwing one teacup at the vehicle and breaking that one cup in the process ("footbridge" scenario).

Mimicking results from research using similar scenarios with people rather than teacups, participants were more likely to say that the child broke a rule in the "footbridge" scenario. However, moral judgments differed depending on whether the scenario was about people or cups. When the scenario was about people, the majority of participants reasoned that it was not okay to violate the rule, but when the scenario was about teacups, the majority of participants reasoned that violating the rule was acceptable (Nichols & Mallon, 2006). The authors interpreted these findings to mean that moral judgments in the case of people are guided by a moral rule against harm ("do not kill innocent people") that does not apply to the teacup case. That is, Nichols & Mallon (2006) interpreted their data in a way consistent with one potential explanation of the data obtained by Lahat et al. (2012), arguing that context (people vs. teacups) may influence moral judgments.

In summary, individuals do not respond inflexibly to distress signals when making moral judgments. Rather, children and adults consider the context of the display as well as relevant rules and social norms governing appropriate responses. Tears alone do not mean that harm has occurred or that moral judgment is required.

5. The role of others' intent in moral judgment

An additional role for context in moral judgment concerns the influence of intent. One interpretation that individuals may make of distress signals is the following: distress signals in a victim do not necessarily indicate that another person intended to harm the victim. That is, person A may have harmed person B, and an observer may interpret this event differently depending on whether the harm was intentional or accidental. Just as participants may reason that distress in response to socially normative behaviors does not necessarily mean that harm has occurred, participants may also use information about actors' intentions to determine the extent to which their actions, including harmful actions, are morally wrong (for evidence that reasoning about intent has ancient evolutionary origins and that this ability can be found among non-human primates, see Call et al., 2004; Phillips et al., 2009).

The study of intent has a rich history in psychology and related fields (see also the discussion in Killen & Rizzo, 2014, this issue). For example, Piaget

(1932) showed that young children claimed that it was worse to accidentally make a large ink stain than to intentionally make a small one, showing that they prioritized outcomes over intentions. Only between the ages of 6 and 10 years did children in Piaget's work begin to prioritize information about intent. Below, we discuss more recent work suggesting that intent may begin to influence moral judgment earlier in development than previously thought.

5.1. The development of 'theory of mind'

Moral judgments require people to be able to reason about the contents of others' minds, including people's intentions and beliefs. The ability to do so is called 'theory of mind'. A standard test of theory of mind — the false belief task — asks children to distinguish their own knowledge from the knowledge of another person. In a classic version of the task, a central character (Sally) places an object in a particular location and then leaves the room, at which point another character (Anne) surreptitiously moves the hidden object to a different location. The experimenter then asks participants where Sally will look for the object when she returns to the room. Toddlers younger than four years old typically respond that Sally will search in the object's current location, despite the fact that Sally had no way of knowing that the object was moved (see Wellman et al., 2001, for a review). Researchers have used such findings to argue that infants and young toddlers do not represent others' minds as different from their own; that is, before reaching four years old, children think that everyone has access to the same knowledge (e.g., Wimmer & Perner, 1983). However, more recent findings (see Baillargeon et al., 2010, for a review) indicate that false belief understanding may emerge in the second year of life, suggesting that even infants may represent others' beliefs, even if those beliefs differ from their own (for a discussion of theory of mind among non-human primates, see Premack & Woodruff, 1978; Heyes, 1998; de Waal & Ferrari, 2012).

5.2. The development of intent-based moral judgments

Supporting the claim made by Baillargeon and colleagues that even infants can reason about others' mental states, a number of experiments have shown that, beginning in infancy, individuals' responses to and moral evaluations of actors depend on the actor's intent. One line of work (Dahl et al., in press) suggests that preferential helping based on intent emerges gradually over the first two years of life. In this study, 17- and 22-month-old infants had the opportunity to help actors who had previously acted pro-socially or anti-socially. Infants helped both actors equally. Two-year-old toddlers preferentially helped the pro-social actor when given a choice between helping the two actors but were willing to help the anti-social actors when the pro-social actor was not present.

In another line of work (Hamlin, 2013), 8-month-old infants preferred characters who intended but failed to help others over characters who intended but failed to harm others. That is, infants preferred characters with good intentions rather than characters associated with good outcomes. Furthermore, infants failed to distinguish between characters who intended but failed to help and characters who helped successfully. Older (21-month-old) infants showed their preferences in their behaviors; they selectively helped actors who, in a previous interaction, intended to provide a toy, regardless of whether the actors succeeded or failed in carrying out their goal to help (Dunfield & Kuhlmeier, 2010). Three-year-old toddlers provided less help to actors who had performed harmful actions in the past or who demonstrated that they had harmful intentions, even in the absence of actual harms (Vaish et al., 2010). And, like adults, four-year-olds judged intentional harms to be worse than accidental harms and showed greater emotional arousal, as measured by pupil dilation, to scenarios depicting intentional rather than accidental harm (Decety et al., 2012).

Intent-based moral judgment continues to develop between the ages of four and eight years (Cushman et al., 2013). Cushman and colleagues used evidence from young children to argue for a two-process model of moral judgment (see also Cushman, 2008, for evidence supporting a similar model in adults). In their study, children heard stories concerning attempted harm (e.g., a boy tried to push over another child but tripped on a rock instead) and unintentional harm (e.g., a boy tripped over a rock while running and accidentally pushed someone over in the process). Participants then delivered moral judgments (e.g., "Should [the character] be punished?", "Is [the character] a bad, naughty boy?"). When collapsing across stories and dependent measures, the researchers found that with increasing age, children became increasingly likely to condemn attempted harm despite the fact that the outcome was benign. Older children were also less likely than younger children to condemn accidental harm. These results show an age-related shift to greater reliance on intent rather than outcome when making moral judgments of others' actions.

Additional effects found in this study shed light on the influence of intent-based moral judgment on judgments concerning punishment (Cushman et al., 2013). Specifically, older children relied more on information about intent when judging the character's 'naughtiness', compared to when judging the extent to which the character should be 'punished'. When responding to stories involving accidental harm, intent-based naughtiness judgments mediated the effect of age on intent-based punishment judgments, but the reverse was not the case. Furthermore, initial intent-based naughtiness judgments led to greater subsequent intent-based punishment judgments, but the reverse did not occur. These findings suggest that intent-based naughtiness judgments constrained intent-based punishment judgments. Furthermore, Cushman et al. (2013) use these results to argue in favor of the idea that the developmental shift from outcome-based reasoning to intent-based reasoning relies on conceptual changes within the moral domain rather than gains in more domain-general abilities such as executive function and theory of mind.

Other work, however, has investigated the ways in which the development of theory of mind abilities may influence the development of moral cognition and vice versa. Findings from this literature, in combination with Cushman et al.'s (2013) research, suggest that changes in children's moral judgments may depend both on conceptual change within the domain of morality and on the development of more domain-general abilities. For example, in one study (Killen et al., 2011), 3.5–7.5-year-old children who did not exhibit full competence on a task measuring morally-relevant theory of mind (MoTOM) were more likely to attribute negative intentions to a peer who accidentally harmed another than did participants who answered all MoTOM questions correctly. In a follow-up study, participants who did not pass the MoTOM task reported that it was more acceptable to punish the "accidental transgressor" than did participants who answered all MoTOM questions correctly. These studies point to a relationship between developing moral judgments and the emergence of theory of mind.

Additional evidence suggests that moral judgments may also play a role in influencing theory of mind. For example, Leslie and colleagues (2006) found that preschoolers were more likely to say that a person intentionally caused a negative rather than a positive outcome, despite the fact that both outcomes were presented as unintended. Similar results have been found among adults (Knobe, 2005). The reverse is also true, as demonstrated by evidence showing that moral cognition recruits brain regions that support mental state processing, such as the right temporoparietal junction (RTPJ) and medial prefrontal cortex (MPFC; Kedia et al., 2008; Young & Saxe, 2009). These data suggest that healthy adults reason about others' mental states when delivering moral judgments. Additional neuroscience evidence points to the importance of neurodevelopment for moral judgment (Decety & Howard, 2013). For example, in one study (Decety & Michalska, 2010), 7–40-year-old participants viewed scenarios where individuals experienced either intentional or accidental physical harm. An age-related change was observed in ventro-medial pre-frontal cortex (VMPFC) activation. Whereas younger participants demonstrated activation in the medial VMPFC when exposed to intentional harm, the locus of activation moved to the lateral VMPFC as participants aged. This demonstrates a shift from a more vis-ceral response (typically associated with the medial VMPFC) to a more cognitive response integrating information about mental and affective states (typically associated with the lateral VMPFC). Thus, neurodevelopmental changes may underlie some changes in moral cognition across development.

5.3. Intent-based moral judgments in adulthood

Intent plays such an important role in moral judgment that, in some cases, participants prioritize information about intent rather than outcome when evaluating actions. For example, in one set of studies (Cushman, 2008), adults read vignettes that manipulated the actor's desire (e.g., the actor wanted or did not want to burn another person's hand), the actor's belief (e.g., the actor thought or did not think that her action would burn another person's hand), and the outcome (e.g., the other person's hand was burned or not burned). Adults then judged how morally wrong the actor's behavior was, how much the actor was to blame for the outcome, and how much the actor should be punished. When judging moral wrongness, adults prioritized information about the actor's intent. By contrast, when assessing blame and punishment, adults also considered the harmfulness of the outcome.

Intent appears to be especially important in adults' judgments of harm-ful — as opposed to purity-violating — actions (Young & Saxe, 2011). Young and Saxe presented participants with vignettes that varied in two ways. First, some vignettes described harmful actions (e.g., one person poi-soned another), while others described purity-violating actions that did not cause harm (e.g., two long-lost siblings had consensual sex). Second, within each condition, some vignettes described actors who behaved intentionally or accidentally (e.g., the person knew or did not know she was poisoning another person's food; sexual partners knew or did not know they were siblings). Participants judged intentional harmful actions as well as intentional purity-violating actions to be wrong, showing that adults make moral judgments even in the absence of harm. Two additional results of particular relevance to the role of intent in moral judgment emerged. First, participants judged harmful actions to be morally worse when committed intentionally versus accidentally, showing that most adults care about an actor's intent and not just the action's outcome when determining moral wrongness. Second, accidental harms were judged less morally wrong than accidental purity violations. Adults did not rely on intent indiscriminately when making moral judgments; rather, information about intent mattered more for judgments concerning harm than for judgments concerning purity. These results may have occurred because harmful actions usually impact a victim, while purity-violating actions do not need to impact anyone other than the perpetrators (see also Chakroff et al., in press).

In summary, moral judgment does not depend solely on harmful outcomes. Rather, people demonstrate a sophisticated ability to consider actors' intentions as well as outcomes for moral judgment. Additionally, individuals deliver moral judgments even in cases (e.g., purity violations) where no victims appear to be harmed (see also Haidt, 2001, 2012; Graham et al., 2011; Koleva et al., 2012). As children mature, they become better able to process mental state information for moral judgment.

6. Connections between moral cognition and other domains

We have already discussed work showing that young children and adults distinguish moral rules from other types of norms, such as norms governing social convention. In addition to distinguishing morality from social convention, children and adults also distinguish morality from mere preference. Adults place morality in an intermediate position between beliefs about facts on the one hand and preferences on the other; the logic is that morality is similar to objective fact in some ways and similar to subjective preference in other ways (Goodwin & Darley, 2008). Preschoolers also treat moral properties like "good" and "bad" as more objective than properties that depend more on preference, such as "fun" and "icky" (Nichols & Folds-Bennett, 2003). Of course, moral beliefs cannot be verified in the same way that facts can be identified as true or false. For example, the factual statement "George Washington was the first president of the United States" can be verified using the proper materials (e.g., textbooks, original documents) and skills (e.g., reading). By contrast, the moral belief that "hitting is wrong" cannot be verified in the same way.

A discussion of moral objectivity is beyond the scope of this article (for further discussion, see Nichols, 2004; Wainryb et al., 2004; Sarkissian et al., 2011; Goodwin & Darley, 2012; Young & Durwin, 2013), but it is important to recognize that morals beliefs are not the only beliefs that occupy this intermediate space. Children between five and ten years old, as well as adults, place religious beliefs in the same intermediate space (Heiphetz et al., 2013). Adolescents and adults treat political beliefs similarly, positioning them between factual and preference-based beliefs (Heiphetz et al., data not shown). Although children and adults distinguish morality from some types of beliefs, such as those concerning preference and social convention, they may group morality with other ideologies, including religion and politics. These apparently different domains — morality, religion, and politics — may share common psychological processes. For example, all concern ideologies — shared beliefs about how the world is and, importantly, how the world ought to be (Jost et al., 2009). Future work should investigate the cognitive signatures of ideological thought.

7. Conclusion

From infancy to adulthood, people make sophisticated moral judgments that rely on a number of inputs, such as distress signals, social norms, moral rules, and information about an actor's intent. First, we discussed evidence indicating that distress signals can indicate that harm has occurred. Second, we presented work showing that moral judgment often occurs in the absence of distress signals. Third, we presented research showing that children and adults alike distinguish harmful actions from violations of social convention and that social norms as well as moral rules can influence moral judgments. In addition, this body of research shows that even toddlers may withhold moral judgment in the presence of distress signals if they perceive the distress to occur as a result of a socially normative (rather than harmful) behavior. Fourth, children and adults alike use information about others' intentions to inform their moral judgments. Finally, although individuals also distinguish moral beliefs from beliefs about facts and preferences, they appear to group morality with other ideologies such as political and religious beliefs. Exploring further connections between moral cognition and other domains remains a fruitful avenue for future research.

Acknowledgements

The authors wish to thank J. Kiley Hamlin for helpful discussion. The first author was supported by a National Academy of Education/Spencer Dissertation Completion Fellowship while working on this project.

References

Ainsworth, M.D.S., Blehar, M.C., Waters, E. & Wall, S. (1978). Patterns of attachment: a psychological study of the strange situation. — Lawrence Erlbaum, Hillsdale, NJ.

Baillargeon, R., Scott, R.M. & He, Z. (2010). False-belief understanding in infants. — Trends Cogn. Sci. 14: 110-118.

Bekoff, M. (2001). Social play behaviour: cooperation, fariness, trust, and the evolution of morality. — J. Consciousness Stud. 8: 81-90.

Blair, R.J.R. (1995). A cognitive developmental approach to morality: investigating the psychopath. — Cognition 57: 1-29.

Blair, R.J.R. & Morton, J. (1995). Putting cognition into psychopathy. — Behav. Brain Sci. 18: 431-437.

Boehm, C. (2014). The moral consequences of social selection. — Behaviour 151: 167-183.

Call, J., Hare, B., Carpenter, M. & Tomasello, M. (2004). 'Unwilling' versus 'unable': chimpanzees' understanding of human intentional action. — Dev. Sci. 7: 488-498.

Chakroff, A., Dungan, J. & Young, L. (in press). Harming ourselves and defiling others: what determines a moral domain? — PLoS One.

Côté, S., Piff, P.K. & Willer, R. (2013). For whom do the ends justify the means? Social class and utilitarian moral judgment. — J. Pers. Soc. Psychol. 104: 490-503.

Cushman, F. (2008). Crime and punishment: distinguishing the roles of causal and intentional analyses in moral judgment. — Cognition 108: 353-380.

Cushman, F., Gray, K., Gaffey, A. & Mendes, W.B. (2012). Simulating murder: the aversion to harmful action. — Emotion 12: 2-7.

Cushman, F., Sheketoff, R., Wharton, S. & Carey, S. (2013). The development of intent-based moral judgment. — Cognition 127: 6-21.

Dahl, A., Schuck, R.K. & Campos, J.J. (in press). Do young toddlers act on their social preferences? — Dev. Psychol., DOI:10.1037/a0031460

de Waal, F.B.M. (1993). Reconciliation among primates: a review of empirical evidence and unresolved issues. — In: Primate social conflict (Mason, W.A. & Mendoza, S.P., eds). State University of New York Press, Albany, NY, p. 111-144.

de Waal, F.B.M. (2014). Natural normativity: the "is" and "ought" of animal behavior. — Behaviour 151: 185-204.

de Waal, F.B.M. & Ferrari, P.F. (eds) (2012). The primate mind: built to connect with other minds. — Harvard University Press, Cambridge, MA.

Decety, J. & Howard, L.H. (2013). The role of affect in the neurodevelopment of morality. — Child Dev. Perspect. 7: 49-54.

Decety, J. & Michalska, K.J. (2010). Neurodevelopmental changes in the circuits underlying empathy and sympathy from childhood to adulthood. — Dev. Sci. 13: 886-899.

Decety, J., Michalska, K.J. & Kinzler, K.D. (2012). The contribution of emotion and cognition to moral sensitivity: a neurodevelopmental study. — Cereb. Cortex 22: 209-220.

den Bak, I.M. & Ross, H.S. (1996). I'm telling! The content, context, and consequences of children's tattling on their siblings. — Soc. Dev. 5: 292-309.

Dunfield, K.A. & Kuhlmeier, V.A. (2010). Intention-mediated selective helping in infancy. — Psychol. Sci. 21: 523-527.

Dunfield, K.A., Kuhlmeier, V.A., O'Connell, L. & Kelley, E. (2011). Examining the diversity of prosocial behavior: helping, sharing, and comforting in infancy. — Infancy 16: 227-247.

Eibl-Eibesfeldt, I. (1970). Ethology: the biology of behavior. — Holt, Rinehart & Winston, Oxford.

Goodwin, G.P. & Darley, J.M. (2008). The psychology of meta-ethics: exploring objectivism. — Cognition 106: 1339-1366.

Goodwin, G.P. & Darley, J.M. (2012). Why are some moral beliefs perceived to be more objective than others? — J. Exp. Soc. Psychol. 48: 250-256.

Graham, J., Haidt, J. & Nosek, B.A. (2009). Liberals and conservatives rely on different sets of moral foundations. — J. Pers. Soc. Psychol. 96: 1029-1046.

Graham, J., Nosek, B.A., Haidt, J., Iyer, R., Koleva, S. & Ditto, P.H. (2011). Mapping the moral domain. — J. Pers. Soc. Psychol. 101: 366-385.

Greene, J.D., Sommerville, R.B., Nystrom, L.E., Darley, J.M. & Cohen, J.D. (2001). An fMRI investigation of emotional engagement in moral judgment. — Science 293: 2105-2108.

Haidt, J. (2001). The emotional dog and its rational tail: a social intuitionist approach to moral judgment. — Psychol. Rev. 108: 814-834.

Haidt, J. (2012). The righteous mind: why good people are divided by politics and religion. — Pantheon/Random House, New York, NY.

Hamlin, J.K. (2012). A developmental perspective on the moral dyad. — Psychol. Inquiry 23: 166-171.

Hamlin, J.K. (2013). Failed attempts to help and harm: intention versus outcome in preverbal infants' social evaluations. — Cognition 128: 451-474.

Hamlin, J.K. & Wynn, K. (2011). Young infants prefer prosocial to antisocial others. — Cogn. Dev. 26: 30-39.

Hamlin, J.K., Wynn, K. & Bloom, P. (2007). Social evaluation in preverbal infants. — Nature 450: 557-559.

Hamlin, J.K., Wynn, K. & Bloom, P. (2010). Three-month-olds show a negativity bias in their social evaluations. — Dev. Sci. 13: 923-929.

Heiphetz, L., Spelke, E.S., Harris, P.L. & Banaji, M.R. (2013). The development of reasoning about beliefs: fact, preference, and ideology. — J. Exp. Soc. Psychol. 49: 559-565.

Hepach, R., Vaish, A. & Tomasello, M. (2013). Young children sympathize less in response to unjustified emotional distress. — Dev. Psychol. 49: 1132-1138.

Heyes, C.M. (1998). Theory of mind in nonhuman primates. — Behav. Brain Sci. 21: 101-134.

Hume, D. (1739/2012). A treatise of human nature. — Available online at: http://www.gutenberg.org/files/4705/4705-h/4705-h.htm.

Ingram, G.P.D. & Bering, J.M. (2010). Children's tattling: the reporting of everyday norm violations in preschool settings. — Child Dev. 81: 945-957.

Jost, J.T., Federico, C.M. & Napier, J.L. (2009). Political ideology: its structure, functions, and elective affinities. — Ann. Rev. Psychol. 60: 307-337.

Joyce, R. (2014). The origins of moral judgment. — Behaviour 151: 261-278.

Kedia, G., Berthoz, S., Wessa, M., Hilton, D. & Martinot, J. (2008). An agent harms a victim: a functional magnetic resonance imaging study on specific moral emotions. — J. Cogn. Neurosci. 20: 1788-1798.

Killen, M. & Rizzo, M. (2014). Morality, intentionality, and intergroup attitudes. — Behaviour 151: 337-359.

Killen, M., Mulvey, K.L., Richardson, C., Jampol, N. & Woodward, A. (2011). The accidental transgressor: morally-relevant theory of mind. — Cognition 119: 197-215.

Knobe, J. (2005). Theory of mind and moral cognition: exploring the connections. — Trends Cogn. Sci. 9: 357-359.

Kohlberg, L. (1969). Stage and sequence: the cognitive developmental approach to socialization. — In: Handbook of socialization theory and research (Goslin, D.A., ed.). Rand McNally, Chicago, IL, p. 347-380.

Koleva, S.P., Graham, J., Iyer, R., Ditto, P.H. & Haidt, J. (2012). Tracing the threads: how five moral concerns (especially Purity) help explain culture war attitudes. — J. Res. Pers. 46: 184-194.

Lahat, A., Helwig, C.C. & Zelazo, P.D. (2012). Age-related changes in cognitive processing of moral and social conventional violations. — Cogn. Dev. 27: 181-194.

Leslie, A.M., Knobe, J. & Cohen, A. (2006). Acting intentionally and the side-effect effect: theory of mind and moral judgment. — Psychol. Sci. 17: 421-427.

Leslie, A.M., Mallon, R. & DiCorcia, J.A. (2006). Transgressors, victims, and cry babies: is basic moral judgment spared in autism? — Soc. Neurol. 1: 270-283.

Lockhart, K.L., Abrahams, B. & Osherson, D.N. (1977). Children's understanding of uniformity in the environment. — Child Dev. 48: 1521-1531.

Lorenz, K. (1966). On aggression. — Bantam Books, New York, NY.

Martin, G.B. & Clark, R.D. (1982). Distress crying in neonates: species and peer specificity. — Dev. Psychol. 18: 3-9.

Nichols, S. (2004). After objectivity: an empirical study of moral judgment. — Philos. Psychol. 17: 3-26.

Nichols, S. & Folds-Bennett, T. (2003). Are children moral objectivists? Children's judgments about moral and response-dependent properties. — Cognition 90: B23-B32.

Nichols, S. & Mallon, R. (2006). Moral dilemmas and moral rules. — Cognition 100: 530-542.

Nichols, S.R., Svetlova, M. & Brownell, C.A. (2009). The role of social understanding and empathic disposition in young children's responsiveness to distress in parents and peers. — Cogn. Brain Behav. 13: 449-478.

Nucci, L. (1981). Conceptions of personal issues: a domain distinct from moral or societal concepts. — Child Dev. 52: 114-121.

Phillips, W., Barnes, J.L., Mahajan, N., Yamaguchi, M. & Santos, L. (2009). 'Unwilling' versus 'unable': capuchin monkeys' (*Cebus paella*) understanding of human intentional action. — Dev. Sci. 12: 938-945.

Piaget, J. (1932). The moral judgment of the child. — Harcourt, Oxford.

Premack, D. & Woodruff, G. (1978). Does the chimpanzee have a theory of mind? — Behav. Brain Sci. 1: 515-526.

Quinn, W.S. (1989). Actions, intentions, and consequences: the doctrine of double effect. — Philos. Publ. Aff. 18: 334-351.

Rakoczy, H. (2008). Taking fiction seriously: young children understand the normative structure of joint pretense games. — Dev. Psychol. 44: 1195-1201.

Rakoczy, H., Warneken, F. & Tomasello, M. (2008). The sources of normativity: young children's awareness of the normative structure of games. — Dev. Psychol. 44: 875-881.

Rakoczy, H., Brosche, N., Warneken, F. & Tomasello, M. (2009). Young children's understanding of the context-relativity of normative rules in conventional games. — Br. J. Dev. Psychol. 27: 445-456.

Romero, T. & de Waal, F.B.M. (2010). Chimpanzee (*Pan troglodytes*) consolation: third-party identity as a window on possible function. — J. Comp. Psychol. 124: 278-286.

Romero, T., Castellanos, M.A. & de Waal, F.B.M. (2010). Consolation as possible expression of sympathetic concern among chimpanzees. — Proc. Natl. Acad. Sci. USA 107: 12110-12115.

Ross, H.S. & den Bak-Lammers, I.M. (1998). Consistency and change in children's tattling on their siblings: children's perspectives on the moral rules and procedures in family life. — Soc. Dev. 7: 275-300.

Sagi, A. & Hoffman, M.L. (1976). Empathic distress in the newborn. — Dev. Psychol. 12: 175-176.

Sarkissian, H., Park, J., Tien, D., Wright, J.C. & Knobe, J. (2011). Folk moral relativism. — Mind Lang. 26: 482-505.

Schmidt, M.F.H., Rakoczy, H. & Tomasello, M. (2011). Young children attribute normativity to novel actions without pedagogy or normative language. — Dev. Sci. 14: 530-539.

Schmidt, M.F.H., Rakoczy, H. & Tomasello, M. (2012). Young children enforce social norms selectively depending on the violator's group affiliation. — Cognition 124: 325-333.

Shallow, C., Iliev, R. & Medin, D. (2011). Trolley problems in context. — J. Decision Making 6: 593-601.

Sheskin, M. & Santos, L. (2012). The evolution of morality: which aspects of human moral concerns are shared with nonhuman primates? — In: The Oxford handbook of comparative evolutionary psychology (Vonk, J. & Shackelford, T.K., eds). Oxford University Press, New York, NY, p. 434-450.

Smetana, J.G. (1981). Preschool children's conceptions of moral and social rules. — Child Dev. 52: 1333-1336.

Smetana, J.G. (2006). Social-cognitive domain theory: consistencies and variations in children's moral and social judgment. — In: Handbook of moral development (Killen, M. & Smetana, J.G., eds). Lawrence Erlbaum, Mahwah, p. 119-153.

Smetana, J.G., Schlagman, N. & Adams, P.W. (1993). Preschool children's judgments about hypothetical and actual transgressions. — Child Dev. 64: 202-214.

Smith, C.E., Blake, P.R. & Harris, P.L. (2013). I should but I won't: why young children endorse norms of fair sharing but do not follow them. — PLoS One 8: e59510.

Thomson, J.J. (1976). Killing, letting die, and the trolley problem. — Monist 59: 204-217.

Turiel, E. (1983). The development of social knowledge: morality and convention. — Cambridge University Press, Cambridge.

Vaish, A., Carpenter, M. & Tomasello, M. (2009). Sympathy through affective perspective taking and its relation to prosocial behavior in toddlers. — Dev. Psychol. 45: 534-543.

Vaish, A., Carpenter, M. & Tomasello, M. (2010). Young children selectively avoid helping people with harmful intentions. — Child Dev. 81: 1661-1669.

Vaish, A., Missana, M. & Tomasello, M. (2011). Three-year-old children intervene in third-party moral transgressions. — Brit. J. Dev. Psychol. 29: 124-130.

Wainryb, C., Shaw, L.A., Langley, M., Cottam, K. & Lewis, R. (2004). Children's thinking about diversity of belief in the early school years: judgments of relativism, tolerance, and disagreeing persons. — Child Dev. 75: 687-703.

Wellman, H.M., Cross, D. & Watson, J. (2001). Meta-analysis of theory of mind development: the truth about false belief. — Child Dev. 72: 655-684.

Wimmer, H. & Perner, J. (1983). Beliefs about beliefs: representation and constraining function of wrong beliefs in young children's understanding of deception. — Cognition 13: 103-128.

Wyman, E., Rakoczy, H. & Tomasello, M. (2009). Normativity and context in young children's pretend play. — Cogn. Dev. 24: 146-155.

Young, L. & Durwin, A.J. (2013). Moral realism as moral motivation: the impact of meta-ethics on everyday decision-making. — J. Exp. Soc. Psychol. 49: 302-306.

Young, L. & Saxe, R. (2009). An fMRI investigation of spontaneous mental state inference for moral judgment. — J. Cogn. Neurosci. 21: 1396-1405.

Young, L. & Saxe, R. (2011). When ignorance is no excuse: different roles for intent across moral domains. — Cognition 120: 202-214.

Zahn-Waxler, C., Radke-Yarrow, M., Wagner, E. & Chapman, M. (1992). Development of concern for others. — Dev. Psychol. 28: 126-136.

Review

Morality, intentionality and intergroup attitudes

Melanie Killen [*] **and Michael T. Rizzo**

Dept. of Human Development and Quantitative Methodology, University of Maryland,
3304 Benjamin Building, College Park, MD 20814, USA
[*]Corresponding author's e-mail address: mkillen@umd.edu

Accepted 22 August 2013; published online 5 December 2013

Abstract

Morality is at the core of what it means to be social. Moral judgments require the recognition of intentionality, that is, an attribution of the target's intentions towards another. Most research on the origins of morality has focused on intragroup morality, which involves applying morality to individuals in one's own group. Yet, increasingly, there has been new evidence that beginning early in development, children are able to apply moral concepts to members of an outgroup as well, and that this ability appears to be complex. The challenges associated with applying moral judgments to members of outgroups includes understanding group dynamics, the intentions of others who are different from the self, and having the capacity to challenge stereotypic expectations of others who are different from the ingroup. Research with children provides a window into the complexities of moral judgment and raises new questions, which are ripe for investigations into the evolutionary basis of morality.

Keywords

moral judgment, developmental psychology, intergroup relations, social exclusion.

1. The origins of morality

Developmental psychologists have demonstrated that children resolve conflicts in non-aggressive ways. Through negotiation, bargaining, and compromising about the exchange of resources, for example, children learn about reciprocity, mutuality, equality and fairness (Killen & Rutland, 2011). Parallel findings have been revealed with non-human primates in which animals resolve conflicts in ways other than through aggressive means (Cords & Killen, 1998; Verbeek et al., 2000; de Waal, 2006). These findings point to a social origin of nature, rejecting a solely aggressive view of human or non-human primate nature. In developmental psychology, theories of the origins of sociality stem from biological and evolutionary theories (de Waal, 1996, 2006), as well as from philosophical theories for defining morality (Rawls, 1971; Gewirth, 1978; Sen, 2009). Fundamentally, a developmental approach is one that examines the ontogenetic course of sociality, studying the change over time, origins, emergence, and source of influence.

In this article, we review the developmental evidence for the emergence of sociality, the origins of moral judgments, and the developmental trajectories regarding moral judgment through childhood. We identify the social interactional basis for moral judgments and the challenges that come with applying moral judgments to everyday interactions. A basic question often posed is whether sociality is learned or innate. Our argument is that it is both. Sociality (and morality) is constructed through a process of interactions, reflections, and evaluations. Peer interactions play a unique role enabling children to engage with equal partners that create conflict which force individuals to reflect, abstract, and form evaluations of their everyday interactions. The argument that sociality precedes morality has also received support from biological and evolutionary accounts of morality (see de Waal, 2014, this issue). This view stems from Piaget's (1932) foundational work on moral judgment in which he observed children playing games and interviewed them about their interpretations of the fairness of the rules that governed their interactions (Carpendale & Lewis, 2006). Piaget's theory was quite different from Freud's (1930/1961) and Skinner's (1974) who postulated that morality was acquired through external socialization, by parents or external contingencies in the environment. Thus, the proposal by Piaget is bottom-up, not top-down. Through a reciprocal process of judgment — action-reflection, children construct knowledge and develop concepts that enable them to function in the world. Adaptation is a key mechanism by which children develop, change, and acquire knowledge. This view of morality is consistent with evolutionary, biological, and comparative perspectives as well (de Waal, 2006; Churchland, 2011). As stated by Darwin (1871/2004), morality is a result of both social predispositions as well as intellectual faculties, defined as reciprocity, or the Golden Rule.

Thus, moral judgments do not emerge in a vacuum. Morality originates through the daily interactions the child experiences with peers, adults, family members, and friends. What happens very early in development is the emergence of group identity, and group affiliation. Through interacting with members of one's own group and other groups, children have the social-cognitive task of balancing individual needs, group needs, and the motivation to be fair and just. What makes it complex is that these challenges involve attributions of intentions of others, both members of the 'ingroup' as well as of the 'outgroup' (as explained below). Intentionality has been studied by moral theorists as well as researchers who examine children's understanding of mental states, often referred to as theory of mind.

The set of challenges created by acquiring moral norms in group interactions, along with a developing sense of intentionality understanding has been studied by both social and developmental psychologists for over a decade (see Mulvey et al., 2013), and this research will constitute a central focus of the article. To do this we draw on several theories: (1) social domain theory (Turiel, 1983; Smetana, 2006) which has identified three domains of knowledge: the moral, societal and psychological (see Figure 1); (2) developmental theories of group identity (Nesdale, 2004; Abrams & Rutland, 2008), which have characterized how group identity emerges in development; and (3) theory of mind (Wimmer & Perner, 1983), which has revealed how children form an understanding of mental states and intentionality. Beginning with social domain theory as the core approach, we have depicted how other dimensions of social life bear on the acquisition of morality (see Figure 1).

Research on moral knowledge has shown that children distinguish moral knowledge from other non-moral social rules early in development, and that their social interactional experiences lead to these judgments. The moral do-

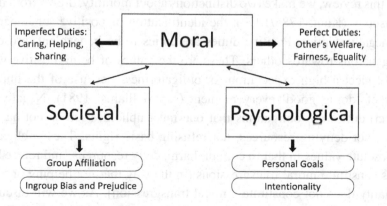

Figure 1. Depiction of three domains of social life, the moral, societal, and psychological, and key constructs investigated in developmental research on morality.

main refers to inter-individual treatment of others with respect to fairness, others' welfare, equal treatment, and justice. In this article, we differentiate morality that pertains to imperfect duties (helping, caring, sharing) from morality that pertains to perfect duties (avoid harm to others, unfair distribution of resources, inequality), which will be discussed in more detail below. The societal domain refers to behavioral uniformities that promote the smooth functioning of social groups, such as customs, conventions, rules, and rituals established by groups. The psychological domain refers to areas of individual prerogatives and personal choice, such as choice of friendships or personal goals. Over 30 years of research has empirically demonstrated that very young children, prior to direct teaching, make distinctions between rules that concern fairness, and those that concern conventions (for example, Turiel, 1998; Smetana, 2011), and that children recognize the psychological domain as an area that is distinct from rules about morality and conventions.

Yet, most conflicts in social life are multifaceted, reflecting multiple domains of knowledge. We are specifically interested in conflicts that reflect the demands to be fair (morality) and to be loyal to the group (group identity) as well as between concerns for fairness, and a lack of the recognition of other's intentions, motivations, and desires (theory of mind). These conflicts reflect everyday interactions between individuals, and in this paper we will highlight these types of experiences which, when resolved, have the potential to contribute to sophisticated moral perspectives.

In this review, we make two distinctions about morality, drawn from moral philosophy (Kant, 1785/1959): the identification of positive (or imperfect) and negative duties. Positive duties are duties to be benevolent, prosocial, and generous towards others. These are the values of being positive which do not meet a high expectation for obligatoriness because of the impossibility of 'doing good' every moment (see Williams, 1981). Negative (or perfect) duties are obligations that one must uphold, such as not harming others, not denying resources, nor refusing basic rights. These obligations are absolute values in that to commit harm, deny resources, and refuse basic rights constitute moral transgressions (in the way that not helping or giving to charity does not constitute a moral transgression). As we will discuss in this article, however, the term 'absolute' takes a different meaning in the context of developmental science research from how it has been formulated in moral philosophy. Developmental science research has revealed the importance of context for determining what creates harm to another, for example, who 'counts' as the 'other', and many other factors, and to this extent the absolute quality of moral norms is not a rigid application of the maxim to all social encounters. The absolute quality of moral norms has to do with the requirement that moral norms are not arbitrary, alterable, changeable, or a matter of group consensus and authority mandates, in contrast to non-moral transgressions (such as conventions or etiquette), as we discuss below.

Much of the very early research on the roots of morality pertains to positive duties. This focus, by definition, provides the building blocks for morality. These data have not yet provided a basis for demonstrating the obligatoriness of morality, which is revealed more clearly when an act constitutes a moral transgression, as distinct from a positive act towards others. As we will review, by childhood, it becomes clear how morality constitutes an understanding of the negative duties. We will review evidence for the emergence of a prosocial disposition in early childhood, and then review research demonstrating moral judgments, and moral orientations in childhood through adolescence. We will conclude with a reflection about cross-species comparisons and humanity in the area of morality.

2. Origins of sociality

By the time infants reach their first birthday, they are already equipped with some of the rudimentary precursors to social and moral behavior. At a most basic level, they understand who people are, and differentiate between humans and other animate and inanimate objects (Thompson, 2006). Additionally, young infants possess at least some understanding of another's goals and intentions (Woodward, 2008, 2009). From these findings, it is clear that by the first birthday, if not sooner, children are already beginning to understand who people are and, more importantly, what makes them special — their actions, desires, and intentions. Infants are not, however, limited to simply understanding who — and what — humans are, they are also predisposed to orient towards, and interact with, other humans. This early social orientation is the core root of social and moral cognition, as it binds the infant into the social world as an active social agent.

With these early capacities, infants also actively interact with others in social ways to form attachments and relationships. Social exchanges and reciprocity with parents (e.g., attention sharing and imitation) are the building blocks of social cognition, and help foster healthy attachments and relationships with others (Cassidy, 2008). These early attachments are then the first steps in the development of a behavioral system that encourages close proximity and — importantly — frequent interactions with parents, and eventually peers. Through these relationships, infants develop the motivation to be socially involved and develop a more sophisticated understanding of others knowledge, emotions, and desires as well as their own self-awareness and awareness of social-conventions and morality (Killen & Rutland, 2011).

Given that from a very young age humans form deep, meaningful attachments and relationships with those around them, and that infants actively interpret experiences with those early relationships, it should not be surprising that infants begin to prefer interactions that are positive and cooperative over those that are negative and preventative. In a recent line of studies building from previous research on helper/hinderer distinctions (Kuhlmeier et al., 2003), Hamlin and colleagues (2007) tested whether infants differentiate between characters who had acted prosocially or antisocially toward another character. Infants as young as 6-months-old were shown a short play on a computer animated screen with geometric shapes as agents. In the play, an agent unsuccessfully attempted to climb a hill, and was then either helped (pushed up the hill) or hindered (pushed down the hill) by another agent. Infants showed a looking time and reaching preference for the helping agent over the hindering agent. Furthermore, they found that infants did not only prefer the helper to the hinderer, but also preferred the helper over a neutral agent (showing a 'liking' for the helper) and preferred a neutral agent over the hinderer (showing an avoidance of the hinderer). This suggests that infants are, indeed, making social judgments based on their observations of others' actions — preferring those who act prosocially and avoiding those who harm others.

While these early social judgments may represent the building blocks to moral judgments, it is not clear to what extent these social judgments are moral judgments themselves. These social judgments may be closer to an "I don't like those who harm others, and I like it when those I don't like are harmed" type of judgment than an 'ought' to type of moral judgment; "One ought not to harm others, and those who harm others ought to be punished". The former type of evaluation is subject to the personal likes and dislikes of the individual infant and may be more in line with personal judgments like,

"I like jazz musicians and dislike rock-and-roll musicians, and I like it when jazz musicians succeed and rock-and-roll musicians fail", whereas the latter focuses on the intrinsic moral rights that extend to all humans.

This line of research has provided valuable insight into infants' social understanding. Infants are able to make social judgments of puppets and geometric shapes in a variety of scenarios, and incorporate the mental states of the agents' into their judgments. Furthermore, infants appear to be making, 'enemy of my enemy' judgments, as they prefer those who punish antisocial agents to those who reward them. From this, it seems clear that infants are not only bound into their social world through their social relationships and attachments, but are evaluating others based on their actions and forming judgments about what they would like to happen to those others.

3. Beneficence

Given that infants as young as 6-months-old prefer agents who help others, an important question is whether infants themselves are willing to help or not, and what guidelines they use to determine who and when to help. Recent developmental research has found that, across a wide range of studies, toddlers and children show the capacity, and tendency, to instrumentally help others achieve their goals. Studies have found that if an experimenter drops an object onto the floor and unsuccessfully tries to reach for it, toddlers 14-months-old and older will instrumentally help the experimenter by walking over to pick up the dropped object and handing it to the experimenter (Warneken & Tomasello, 2006; Over & Carpenter, 2009). Additionally, these prosocial helping behaviors do not appear to be motivated by the expectation of a material reward (Warneken et al., 2007), which has actually been found to decrease helping behaviors after toddlers stopped receiving the reward (Warneken & Tomasello, 2008).

A recent study conducted by Dahl and colleagues (2013) examined if 17-, 22- and 26-month-old toddlers were willing to help an antisocial agent. They found that it was not until 26-months-old that toddlers began to consistently help a prosocial agent over an antisocial agent, and that even 26-month-old toddlers helped an antisocial agent if given the opportunity to. It is important to note, however, that in this study, contrary to the findings of Hamlin and colleagues (2007) with geometric shapes and puppets, it was not until 26-months-old that toddlers distinguished between the prosocial and antisocial, *human* agents while they were acting prosocially or antisocially.

Brownell and colleagues have also examined infants' ability to coordinate their actions in order to cooperate with peers (Brownell & Carriger, 1991; Brownell et al., 2006) which differs from research on cooperation between an infant (14 months) and an informed adult experimenter (see Warneken & Tomasello, 2007). In non-verbal tasks — designed to enable children to co-ordinate their behavior with one another — by Brownell and colleagues, 12- and 18-month-olds were unable to intentionally coordinate their behaviors to achieve a common goal. For the few instances in which 18-month-olds did coordinate their behaviors, it was coincidental and unstructured. No 12- month-old dyad ever cooperated. However, by 24-months-old, toddlers were able to achieve their collective goal through cooperatively coordinating their behaviors and by 30-months-old even gesturally and verbally communicated directions to one another. Brownell and colleagues argued that this develop-mental shift is linked to infants' and toddlers' developing ability to interpret the desires, intentions, and goals of their peers, whose desires, intentions, and goals are especially capricious and difficult to interpret.

The collective findings of the studies presented suggest that infants are deeply involved in their social world through their relationships and attach-ments, make social judgments about others, and, by toddlerhood, are highly motivated to help others when they need assistance to achieve their goal. In-fants begin to establish attachments in the first few months of life. Around 6-months-old, infants begin to make rudimentary social judgments about non-human agents. As toddlers become better apt to move around in their social world, toddlers as young as 14-months-old begin to instrumentally help adults with their physical goals, and begin coordinating their behavior with peers by 24-months-old. Then, around 26-months old, toddlers begin to preferentially help certain people over others.

4. Social interactions and judgments regarding moral and conventional transgressions

During the preschool period moral judgments are spontaneously articulated by children towards others during their encounters (and conflicts). To ex-amine young children's actual responses to one another in social interactions regarding morally relevant issues, Nucci & Turiel (1978) observed children's social interactions during free-play in preschools. Nucci & Turiel (1978) ex-amined the nature of preschool children's and teacher's responses to both conventional and moral transgressions committed by preschool children. They found that both children and adults responded to moral transgressions by focusing primarily on the intrinsic nature of the actions. Children, often the victim of the transgressions, responded with direct feedback regarding the harm or loss they experienced due to the transgression.

Complementing these responses, teachers responded by focusing on the effects of the transgression on the victim when discussing the transgression with the transgressor. However, in contrast to the responses to moral transgressions, only teachers — not children — responded consistently to conventional transgressions (children often ignored the transgressions). Teachers were found to respond to conventional transgressions by mentioning school rules, invoking sanctions or punishments, discussing the disruptive consequences of the transgressions, and by using commands to stop the transgression (e.g., 'sit down'). These findings demonstrate that children hear different messages about, and respond differently to, social interactions in the contexts of moral and social conventional events.

To extend these findings, Nucci & Nucci (1982) examined school-aged children's developing conceptions of moral and conventional transgressions through both observational and interview methodologies. They found that both children's and teachers' responses differed between conventional events and moral events. Consistent with Nucci & Turiel (1978), they found that children were much less likely to respond to conventional transgressions than moral transgressions, and their responses to moral transgressions revolved around the intrinsic harm or injustice to the victim that was caused by the act. Children of all ages also responded to moral transgressions with more appeals for teacher intervention as well as direct retaliatory acts directed at the transgressor, although the latter response increased with age. Children were found to respond to conventional transgressions increasingly with age, and mention concerns for the social order, such as classroom rules and social norms. Furthermore, during interviews with children about ongoing events, children were found to make the conceptual distinction between the observed moral and conventional transgressions. Participants judged moral transgressions to be wrong regardless of the presence or absence of a rule or social norm, but judged conventional transgressions as wrong only if the rule or social norm existed in their school or classroom. These findings indicate that children differentiate between naturally occurring moral and conventional transgressions in terms of their observed reactions as well as in the interview context. A striking finding regarding these data pertains to the correspondence between behavior and judgment, which is a central issue in moral development. Children's behavior regarding moral transgression reflected their judgments in interviews about hypothetical vignettes that closely matched children's everyday interactions.

Similar to the findings in preschool children, school-aged children were more likely to respond to moral than conventional transgressions, and responded with statements about the harm done to, or loss experienced by, the victim in moral contexts and responded with statements about school or classroom rules and norms in conventional contexts. Extending the findings with preschool children, these findings suggest that as children become older they become more responsive to conventional transgressions, and become more likely to actively maintain both moral and conventional rules/norms.

To examine the ontogenetic roots of children's distinction between moral and conventional transgressions Smetana (1984) observed toddlers' (13- to 40-months old) social interactions in daycare-center classrooms. Consistent with the work in preschool and school-aged children, moral transgressions were found to elicit responses from both caregivers and toddlers, whereas conventional transgressions only elicited a response from the caregivers. Interestingly, caregivers responded more frequently to moral transgressions with younger toddlers than older toddlers, and responded more frequently to conventional transgressions with older toddlers than younger toddlers. Caretakers responded to moral transgressions with a wide range of commands, physical restraint, and attempts to divert attention and the children who were the victims of the transgressions responded by emphasizing the harmful consequences of the transgression. However, caretakers responded to conventional transgressions most frequently with commands to stop the behavior and occasionally mentioned the concerns for social organization, such as rules and social norms.

The results of this study support the claim that moral and conventional events are conceptually unique and distinct in their origins, and that this distinction is evident in toddlers by the second year of life. Caretakers responded with a greater range of responses to moral than conventional transgressions, and provided explanations referencing the intrinsic harm and cost to the victim in contrast to conventional transgressions that were frequently left unexplained. It is also important to note, however, that many everyday social events are multifaceted, combining moral and conventional elements.

In both the home and the school setting, there are also many decisions that do not involve violations of regulations, but are rather personal choices up to the child. In the home context, Nucci & Weber (1995) examined child and maternal responses to personal (i.e., assertions of personal choice by the child), conventional, and moral issues. Mothers were found to vary their responses according to the type of event; personal issues were met with indirect messages such as offering choices, whereas moral and conventional issues were met with direct messages such as rule statements and commands. Mothers also negotiated with children more regarding personal issues than in response to moral and conventional issues. Children's responses followed a similar line, in that children often protested for their choices regarding personal issues, but did not contest mothers' responses to conventional and moral issues. Nucci & Weber argued that interactions regarding personal issues are critical to the child's emerging sense of autonomy and are seen as a legitimate topic for indirect messages and negotiation between the mother and child, but that moral and conventional issues are not seen as relevant to building autonomy and are thus discussed with direct statements and commands.

To extend the findings of Nucci & Weber (1995) to examine the school context, Killen & Smetana (1999) observed preschool classrooms and conducted interviews with children regarding ongoing moral, conventional, and personal events. Killen & Smetana (1999) drew from previous literature to identify activities which it could be expected that adults would provide, or children would assert, personal choices during school, including what to do during activity time, what to eat and who to sit next to during lunch time, and where to sit during circle time (De Vries & Zan, 1994; Smetana & Bitz, 1996).

The findings revealed that social interactions regarding personal issues occurred more frequently in the preschool context than moral or conventional transgressions. This is an indication of the centrality of autonomy assertion during the preschool years (e.g., "I want to sit next to my friend, Sam!"). Further, consistent with the previous research reviewed, both children and teachers responded to moral transgressions, with adults focusing on the intrinsic consequences of events such as concerns for others' rights and welfare. Similarly, teachers, but not children, were found to respond to conventional events, and did so by issuing commands and referring to the social order of the classroom, including school and classroom rules and norms.

As was also shown by Nucci & Weber (1995), adults (teachers) were more directive, issuing more commands and direct statements, regarding moral and conventional issues than personal issues, and were equally as likely to support or negate children's assertions of personal choice. Differences were also found based on the specific context of the event; circle time was found to be a more conventionally structured setting than the other contexts, and moral transgressions were most likely to occur when classroom activities were less structured, such as during activity time. The contextual differences in occurrences of moral, conventional, and personal events between different school settings suggests that children develop an understanding that certain concerns are more relevant than others in each of the classroom settings, depending upon the degree of structure of each setting. Finally, in contrast to Nucci & Weber (1995), adults (teachers) rarely negotiated personal choices with children in the classroom. Teachers may be less inclined than mothers to negotiate these issues with children due to the generally more structured nature of the school than the home setting.

Overall, these social interaction and judgment studies provide evidence that children's interactions are related to their judgments, and that not all rules are treated the same. Children as young as 3 years of age respond to and evaluate moral rules involving a victim (e.g., harming others or taking away resources) differently than rules involving a regulation to make groups work well (e.g., where to play with toys or wearing a smock to paint). Individuals believe that rules involving protection of victims are generalizable principles that are not a matter of consensus or authority; there are times when a group or authority may violate these principles and recognizing the independence of moral principles and group norms is a fundamental distinction that is necessary for successful interactions with other members of a community, a culture, or a society.

Research examining these distinctions has been validated through cross-cultural research in a range of countries, including North and South America (U.S. Canada, El Salvador, Mexico, Brazil, Colombia), Western Europe (U.K., Germany, The Netherlands, Spain), Asia (Korea, Japan, China, Hong Kong), Africa (Nigeria), Australia, and the Middle East (Israel, Jordan, Palestinian Territories): (for reviews, see Turiel, 1998; Helwig, 2006; Wainryb, 2006). In the next section, we review a major psychological requirement for making moral judgments: understanding intentionality and others' mental states.

5. Morality and 'theory of mind'

The psychological understanding of others' mental states can often compli-cate moral judgments. During the preschool years, children are still develop-ing their ability to understand the mental states of others' — referred to as theory of mind (ToM) (Wellman & Liu, 2004). The ability to use intention-ality judgments to make moral decisions has been shown to exist at the neu-rological level (Young & Saxe, 2008) and has only recently been examined at the level of explicit evaluation and judgments in children. To examine the link between ToM development and moral development, Smetana and col-leagues (2012) examined the reciprocal associations between children's de-veloping ToM ability and their developing moral judgments in a longitudinal study. Participants were assessed on both their ToM ability and their moral judgments at three time points over one year (initial interview, 6-month fol-low up, and 12-month follow up), and ranged from 2- to 4-years-old at the initial interview. Moral judgments were assessed using four prototypic moral scenarios (hitting, shoving, teasing, and name calling), which were drawn from previous research. For each moral scenario participants responded to six questions, each assessing one of the core criteria of a moral judgment; nonpermissiblity, authority independence, rule independence, nonalterabil-ity, generalizability and deserved punishment. Theory of mind was assessed using five ToM tasks; diverse desires, diverse beliefs, unexpected contents false beliefs, change of location false belief, and belief emotion.

Different patterns were found for each of the criterion questions, suggest-ing a reciprocal relationship between the development of ToM and moral judgment. Participants who judged the moral transgressions as more wrong independent of authority had a more advanced ToM ability 6-months later, for both wave 1 and wave 2. Similarly, participants who judged the moral transgression as more impermissible had a more advanced ToM ability 6-months later for the second wave. These findings suggest that children's attempts to understand social relationships and events may influence their later ToM ability. Conversely, results indicated that a more sophisticated ToM ability led to less prototypic moral judgments 6-months later for other criterion questions. Participants viewed all of the moral transgressions as nonpermissible, authority and rule independent, nonalterable, generalizably wrong and deserving of punishment. Taken together, these results suggest that ToM and moral judgments develop as bidirectional, transactional processes (Smetana et al., 2012). Early experiences interpreting moral trans-gressions influence the development of ToM and the ability to understand the mental states of others allows for a more complex understanding of moral scenarios.

One aspect of children's attributions of intentionality and moral judgment not addressed by these findings, however, is children's understanding of un-intentional moral transgressions, and their attributions of intentions in these contexts. A recent investigation of intentionality and moral judgment con-ducted by Killen and colleagues (2011) investigated whether 3-, 5-, and 7-year-old children's ($N = 162$) ToM competence was related to their at-tributions of intentions and moral judgments of unintentional transgressions. The investigators presented participants with a series of tasks, including pro-totypic contents and location change false-belief ToM tasks, a prototypic moral transgression (pushing someone off a swing), and a morally relevant theory of mind task (MoToM). The MoToM task was a story vignette in which a child brought a special cupcake in a paper bag to lunch. Then, as the child left for recess, a classroom helper entered the room and threw away the paper bag. Thus, the classroom helper threw away the lunch bag while helping to clean up the room. The MoToM measure assessed morally rel-evant embedded contents false belief (What did X think was in the bag?), evaluation of the act (Did you think it was all right or not all right?), attribu-tion of the accidental transgressors' intentions (Did X think she was doing something all right or not all right?), and punishment (Should X be pun-ished?). Reasoning was also obtained for these assessments (e.g., Why?). Participants who correctly responded that the classroom helper thought there was only trash in the bag passed the MoToM assessment.

Killen and colleagues (2011) found that while children reliably passed the prototypic contents false-belief ToM task by 5-years-old, children did not reliably pass the MoToM task until around 7-years-old. This indicates that the saliency of a moral transgression can influence younger children's social cognition regarding other's mental states. As shown in Figure 2, with age, children's attributions of negative intentions for the accidental transgressor declined; in contrast, children's evaluations of the act did not change with age. This was due to the fact that children around 5 and 6 years brought in the consideration of negligence. While the intentions were not negative, the act was wrong because the actor should have looked in the bag, suggesting that future research is needed to study notions about negligence.

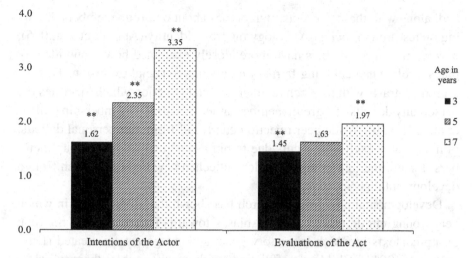

Figure 2. Judgments about intentions and acts in a morally relevant context. Reprinted from Killen et al. (2011).

Additionally, it was found that children who failed the MoToM task were more likely to attribute negative intentions and ascribe punishment to the unintentional transgressor than were children who passed the MoToM task. These finding show that children who do not possess MoToM competency do not understand the unintentional nature of the transgression, and thus attribute negative intentions to, and harsher moral judgments of, transgressors, whereas children who do possess MoToM competency understand that the transgression was unintentional and do not attribute negative intentions to the unintentional transgressor, and do not view the act as wrong.

These findings reveal that as children's understanding of personal/psychological concerns regarding others mental states (intentions) develops, their moral judgments become more complex, in that they can incorporate mental state information into their attributions of intentionality and moral judgments. In the next section, we examine how group identity and intergroup attitudes presents a new challenge for the application of morality to social interactions and social life.

6. Morality and intergroup attitudes

Reasoning about social exclusion involves the moral concerns of fairness and equality, as well as concerns about group functioning, group identity, and, along with these concepts, stereotypes about outgroup members. Drawing on research in social psychology on group identity (Abrams et al., 2005), developmental researchers have more closely examined how group identity plays a role when applying fairness judgments to social exclusion. This research contrasts with research on interpersonal rejection which identifies the personality deficits of a group member rather than group membership (Killen et al., 2013a, b). While peer rejection can result from interpersonal deficits, in many cases exclusion occurs due to biases and dislike of outgroup members. Examining intergroup exclusion reflects a new line of research from a developmental perspective.

Developmental intergroup research has shown the myriad ways in which very young children hold implicit biases towards others, even in minimal group contexts in which laboratory generated categories are created (Dunham et al., 2011). Children's affiliations with groups enables them to hold a group identity which often is reflected by an ingroup preference (Nesdale, 2004). While ingroup preference does not always lead to outgroup dislike, this is often the outcome of an ingroup bias documented in childhood studies on prejudicial attitudes (Bigler et al., 2001). Group identity can be measured many ways (see Rutland et al., 2007), from implicit to explicit forms. Implicit forms refer to when children demonstrate an ingroup bias unbeknownst to them; explicit forms refer to statements in which children articulate a preference for their own group. In a series of studies, researchers have investigated whether children apply moral judgments to group identity preferences, that is, to what extent do they explicitly condone ingroup preference, and when do they view it as unfair?

As one illustration, to investigate how children applied moral concerns about exclusion in stereotypic contexts, Killen and colleagues (2001) studied preschool aged children's ($N = 72$) application of fairness judgments in gender stereotypic contexts. The situations involved a boys' group excluding a girl from truck-playing and a girls' group excluding a boy from doll-playing. There were two conditions for a within-subjects design. In the straightforward condition, children were asked if it was all right to exclude the child who did not fit the stereotype. The majority (87%) of children stated that it was unfair and gave moral reasons ("The boy will feel sad and that's not fair"; "Give the girl a chance to play — they're all the same.").

To create a condition with increased complexity, children were told that two peers wanted to join the group, a boy and a girl, and there was only room for one more to join. Then, they were asked whom should the group pick? In this condition, children were split in their decision, often resorting to stereotypic expectations ("Pick the girl because dolls are for girls."; "Only boys know how to play with trucks."). Children who started out with a fairness judgment did not change their minds after hearing a counter-probe ("But what about the other child?"), whereas children who started out with a stereotypic decision were more likely to change their judgment, and pick the child who did not fit the stereotype. These findings illustrated that children apply moral judgments to contexts involving group identity but often have difficulties when the situation get complicated. Social psychologists find this pattern as well with adults to the extent that adults resort to stereotypic expectations in situations with complexity or ambiguity (Dovidio et al., 2001).

Developmental researchers have also examined exclusion of ingroup members who challenge the norms of the group, and whether this varies by moral or conventional norms (Killen et al., 2013a, b). Children and adolescents evaluated group inclusion and exclusion in the context of norms about resource allocation (moral) and club shirts (social conventional). Participants ($N = 381$), aged 9.5 and 13.5 years, judged an ingroup member's decision to deviate from the norms of the group, whom to include, and whether their personal preference was the same as what they expected a group should do. The findings revealed that participants were favorable towards ingroup members who deviated from unequal group norms to advocate for equal distributions, even when it went against the group norm. Moreover, they were less favorable to ingroup members who advocated for inequality when the group wanted to distribute equally (see Figure 3). The same was true, but less so, for ingroup members who challenged the norms about the club shirts.

The novel findings were that children and adolescents did not evaluate ingroup deviance the same across all contexts; the type of norm mattered, revealing the importance of norms about fairness and equality. In subsequent analyses using the same paradigm, researchers asked children about exclusion of deviant ingroup members, and found that exclusion of deviant members was viewed as increasingly wrong with age. However, participants who disliked deviants found it acceptable to exclude them, such as an ingroup member who wanted to distribute unequally (Hitti et al., in press). Further, Mulvey and colleagues (2012) found that with age, participants differentiated their own favorability from the group's favorability; with age, children expected that groups would dislike members who deviated from the

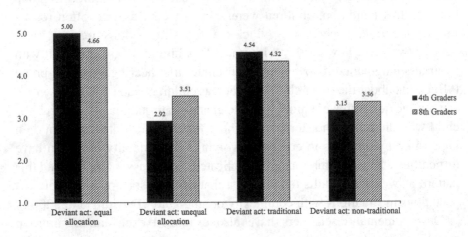

Figure 3. Evaluation of the deviant act by condition and age (1 = really not okay; 6 = really okay). Reprinted from Killen et al. (2013).

group, revealing an aspect of mental state knowledge (that the group would dislike an ingroup member who challenged the group norm even though they would like this member). Overall, the results provide a developmental story about children's complex understanding of group dynamics in the context of moral and social norms. While space does not permit a more exhaustive review of how children apply moral concepts to intergroup attitudes (for a review, see Killen & Rutland, 2011), research from multiple theoretical approaches have investigated how moral judgments and emotions are revealed in children's intergroup attitudes (see Heiphetz & Young, this issue).

7. Conclusions

In summary, the emergence of morality is a long, slow, complex process, which originates early in development and continues throughout life. Infants appear to prefer others who are helpful and have positive intentions towards others. Peer interactions launch the first set of social conflicts that children experience during early childhood. These conflicts, we argue, do not merely reflect a moral versus 'selfish' orientation, but are multifaceted. Children's knowledge about groups and group identity conflict with moral goals of fairness; as well, attributions of intentions curtail children's accurate reading of blame, responsibility and accountability. Further, constructive and positive opportunities to resolve conflicts enables children to develop an understanding of mental states, which along with emerging negotiation and reciprocal social skills, provides the pathway for the development of morality. From childhood to adulthood, moral conflicts become increasingly complicated and multifaceted. Friendships with members of outgroups helps to reduce outgroup bias (Tropp & Prenovost, 2008), and to better understanding the application of intentions to others (Carpendale & Lewis, 2006).

The documentation of sociality reveals the roots for morality. We believe that there is an evolutionary basis to this phenomenon. Non-human primates respond to the distress of another, reject unfair distributions of resources, demonstrate prosocial behavior, and resolve conflicts in non-aggressive ways (de Waal, 2006; Brosnan et al., 2010). This evolutionary past reveals the building blocks for morality in humans. Understanding this complex process enables humans to better reflect on their own morality, and to identify the strategies that appropriately facilitate the development of moral judgment.

Acknowledgements

We thank Frans de Waal, Telmo Pievani and Stefano Parmigiani, the organizers of the 'The evolution of morality: The biology and philosophy of human conscience' workshop, held at the beautiful Ettore Majorana Foundation and Centre for Scientific Culture in Erice, Sicily, June, 2012, for creating a stimulating interdisciplinary context for discussions of morality. We are appreciative of the feedback on the manuscript from Kelly Lynn Mulvey and two anonymous reviewers, and to Shelby Cooley, Laura Elenbaas, Aline Hitti, Kelly Lynn Mulvey and Jeeyoung Noh for collaborations related to the research described in this manuscript. This research was supported by a grant award from the National Science Foundation to the first author.

References

Abrams, D., Hogg, M. & Marques, J. (2005). The social psychology of inclusion and exclusion. — Psychology Press, New York, NY.

Abrams, D. & Rutland, A. (2008). The development of subjective group dynamics. — In: Intergroup relations and attitudes in childhood through adulthood (Levy, S.R. & Killen, M., eds). Oxford University Press, Oxford, p. 47-65.

Bigler, R.S., Brown, C.S. & Markell, M. (2001). When groups are not created equal: effects of group status on the formation of intergroup attitudes in children. — Child Dev. 72: 1151-1162.

Brosnan, S.F., Talbot, C., Ahlgren, M., Lambeth, S.P. & Schapiro, S.J. (2010). Mechanisms underlying responses to inequitable outcomes in chimpanzees, *Pan troglodytes*. — Anim. Behav. 79: 1229-1237.

Brownell, C. & Carriger, M. (1991). Collaborations among toddler peers: individual contributions to social contexts. — In: Perspectives on socially shared cognition (Resnick, L.B., Levine, J.M. & Teasley, S.D., eds). American Psychological Association, Washington, DC, p. 365-383.

Brownell, C.A., Ramani, G.B. & Zerwas, S. (2006). Becoming a social partner with peers: cooperation and social understanding in one- and two-year-olds. — Child Dev. 77: 803-821.

Carpendale, J. & Lewis, C. (2006). How children develop social understanding. — Blackwell Publishing, Oxford.

Cassidy, J. (2008). The nature of the child's ties. — In: Handbook of attachment: theory, research and clinical applications, 2nd edn. (Cassidy, J. & Shaver, P.R., eds). Guilford Publishers, New York, NY, p. 3-22.

Churchland, P. (2011). Braintrust: what neuroscience tells us about morality. — Princeton University Press, Princeton, NJ.

Cords, M. & Killen, M. (1998). Conflict resolution in human and nonhuman primates. — In: Piaget, evolution, and development (Langer, J. & Killen, M., eds). Lawrence Erlbaum Associates, Mahwah, NJ, p. 193-218.

Dahl, A., Schuck, R.K. & Campos, J.J. (2013). Do young toddlers act on their social preferences? — Dev. Psychol., DOI:10.1037/a0031460

Darwin, C. (1871/2004). The descent of man. — Penguin Classics, London.

de Waal, F.B.M. (1996). Good natured: the origins of right and wrong in humans and other animals. — Harvard University Press, Cambridge, MA.

de Waal, F.B.M. (2006). Primates and philosophers: how morality evolved. — Princeton University Press, Princeton, NJ.

de Waal, F.B.M. (2014). Natural normativity: the "is" and "ought" of animal behavior. — Behaviour 151: 185-204.

De Vries, R. & Zan, B. (1994). Moral classrooms, moral children. — Teachers College Press, New York, NY.

Dovidio, J.F., Kawakami, K. & Beach, K. (2001). Implicit and explicit attitudes: examination of the relationship between measures of intergroup bias. — In: Blackwell handbook of social psychology, Vol. 4 (Brown, R. & Gaertner, S.L., eds). Blackwell, Oxford, p. 175-197.

Dunham, Y., Baron, A.S. & Carey, S. (2011). Consequences of 'minimal' group affiliations in children. — Child Dev. 82: 793-811.

Freud, S. (1930/1961). Civilization and its discontents. — W.W. Norton & Company, New York, NY.

Gewirth, A. (1978). Reason and morality. — University of Chicago Press, Chicago, IL.

Hamlin, J.K., Wynn, K. & Bloom, P. (2007). Social evaluation by preverbal infants. — Nature 450: 557-560.

Heiphetz, L. & Young, L. (2014). A social cognitivie developmental perspective on moral judgment. — Behaviour 151: 315-335.

Helwig, C.C. (2006). Rights, civil liberties, and democracy across cultures. — In: Handbook of moral development (Killen, M. & Smetana, J.G., eds). Lawrence Erlbaum Associates, Mahwah, NJ, p. 185-210.

Hitti, A., Mulvey, K.L., Rutland, A., Abrams, D. & Killen, M. (in press). When is it okay to exclude a member of the ingroup?: children's and adolescents' social reasoning. — Soc. Dev.

Kant, I. (1785/1959). Foundations of the metaphysics of morals. — Bobbs-Merrill, New York, NY.

Killen, M. & Rutland, A. (2011). Children and social exclusion: morality, prejudice, and group identity. — Wiley/Blackwell Publishers, New York, NY.

Killen, M. & Smetana, J.G. (1999). Social interactions in preschool classrooms and the development of young children's conceptions of the personal. — Child Dev. 70: 486-501.

Killen, M., Pisacane, K., Lee-Kim, J. & Ardila-Rey, A. (2001). Fairness or stereotypes? Young children's priorities when evaluating group exclusion and inclusion. — Dev. Psychol. 37: 587-596.

Killen, M., Mulvey, K.L., Richardson, C., Jampol, N. & Woodward, A. (2011). The accidental transgressor: morally-relevant theory of mind. — Cognition 119: 197-215.

Killen, M., Mulvey, K.L. & Hitti, A. (2013a). Social exclusion in childhood: a developmental intergroup perspective. — Child Dev. 84: 772-790.

Killen, M., Rutland, A., Abrams, D., Mulvey, K.L. & Hitti, A. (2013b). Development of intra- and intergroup judgments in the context of moral and social-conventional norms. — Child Dev. 84: 1063-1080.

Kuhlmeier, V., Bloom, P. & Wynn, K. (2003). Attribution of dispositional states by 12-month-olds. — Psychol. Sci. 14: 402-408.

Mulvey, K.L., Hitti, A., Cooley, S., Abrams, D., Rutland, A., Ott, J. & Killen, M. (2012). Adolescents' ingroup bias: gender and status differences in adolescents' preference for the ingroup. — Poster presented at the Society for Research Adolescence, Vancouver, BC.

Mulvey, K.L., Hitti, A. & Killen, M. (2013). Morality, intentionality, and exclusion: How children navigate the social world. — In: Navigating the social world: a developmental perspective (Banaji, M. & Gelman, S., eds). Oxford University Press, New York, NY.

Nesdale, D. (2004). Social identity processes and children's ethnic prejudice. — In: The development of the social self (Bennett, M. & Sani, F., eds). Psychology Press, New York, NY, p. 219-245.

Nucci, L.P. & Nucci, M.S. (1982). Children's responses to moral and social-conventional transgressions in free-play settings. — Child Dev. 53: 1337-1342.

Nucci, L.P. & Turiel, E. (1978). Social interactions and the development of social concepts in preschool children. — Child Dev. 49: 400-407.

Nucci, L.P. & Weber, E.K. (1995). Social interactions in the home and the development of young children's conceptions of the personal. — Child Dev. 66: 1438-1452.

Over, H. & Carpenter, M. (2009). Eighteen-month-old infants show increased helping following priming with affiliation. — Psychol. Sci. 20: 1189-1193.

Piaget, J. (1932). The moral judgment of the child. — Free Press, New York, NY.

Rawls, J. (1971). A theory of justice. — Harvard University Press, Cambridge, MA.

Sen, A.K. (2009). The idea of justice. — Harvard University Press, Cambridge, MA.

Rutland, A., Abrams, D. & Levy, S.R. (2007). Extending the conversation: transdisciplinary approaches to social identity and intergroup attitudes in children and adolescents. — Int. J. Behav. Dev. 31: 417-418.

Skinner, B.F. (1974). About behaviorism. — Vintage Books, New York, NY.

Smetana, J.G. (1984). Toddlers' social interactions regarding moral and social transgressions. — Child Dev. 55: 1767-1776.

Smetana, J.G. (2006). Social-cognitive domain theory: consistencies and variations in children's moral and social judgments. — In: Handbook of moral development (Killen, M. & Smetana, J.G., eds). Lawrence Erlbaum Associates, Mahwah, NJ, p. 119-154.

Smetana, J.G. (2011). Adolescents, families, and social development: how teens construct their worlds. — Wiley/Blackwell, New York, NY.

Smetana, J.G. & Bitz, B. (1996). Adolescents' conceptions of teachers' authority and their relations to rule violations in school. — Child Dev. 67: 202-214.

Smetana, J.G., Jambon, M., Conry-Murray, C. & Sturge-Apple, M. (2012). Reciprocal associations between young children's moral judgments and their developing theory of mind. — Dev. Psychol. 48: 1144-1155.

Thompson, R.A. (2006). The development of the person: social understanding, relationships, conscience, self. — In: Handbook of child psychology, Vol. 3: social, emotional, and personality development (Eisenberg, N., ed.). Wiley & Sons, New York, NY, p. 24-98.

Tropp, L.R. & Prenovost, M.A. (2008). The role of intergroup contact in predicting children's inter-ethnic attitudes: evidence from meta-analytic and field studies. — In: Intergroup attitudes and relations in childhood through adulthood (Levy, S.R. & Killen, M., eds). Oxford University Press, Oxford, p. 236-248.

Turiel, E. (1983). The development of social knowledge: morality and convention. — Cambridge University Press, Cambridge.

Turiel, E. (1998). The development of morality. — In: Handbook of child psychology, Vol. 3: social, emotional, and personality development, 5th edn. (Damon, W., ed.). Wiley, New York, NY, p. 863-932.

Verbeek, P., Hartup, W. & Collins, W.A. (2000). Conflict management in children and adolescents. — In: Natural conflict resolution (Aureli, F. & de Waal, F.B.M., eds). University of California Press, Berkeley, CA, p. 34-53.

Wainryb, C. (2006). Moral development in culture: diversity, tolerance, and justice. — In: Handbook of moral development (Killen, M. & Smetana, J.G., eds). Lawrence Erlbaum Associates, Mahwah, NJ, p. 211-240.

Warneken, F., Hare, B., Melis, A., Hanus, D. & Tomasello, M. (2007). Spontaneous altruism by chimpanzees and young children. — PLOS Biol. 5: e184.

Warneken, F. & Tomasello, M. (2006). Altruistic helping in human infants and young chimpanzees. — Science 311: 1301-1303.

Warneken, F. & Tomasello, M. (2008). Extrinsic rewards undermine altruistic tendencies in 20-month-olds. — Dev. Psychol. 44: 1785-1788.

Wellman, H.M. & Liu, D. (2004). Scaling of theory-of-mind tasks. — Child Dev. 75: 502-517.

Williams, B. (1981). Moral luck. — Cambridge University Press, Cambridge.

Wimmer, H. & Perner, J. (1983). Beliefs about beliefs: representation and constraining function of wrong beliefs in young children's understanding of deception. — Cognition 13: 103-128.

Woodward, A.L. (2008). Infants grasp of others intentions. — Curr. Direct. Psychol. Sci. 18: 53-57.

Woodward, A.L. (2009). Infants' learning about intentional action. — Oxford University Press, New York, NY.

Young, L. & Saxe, R. (2008). The neural basis of belief encoding and integration in moral judgment. — Neuroimage 40: 1912-1920.

Section 4: Religion

[When citing this chapter, refer to Behaviour 151 (2014) 363–364]

Introduction

What are the connections in human evolution between supernatural beliefs, organized religion, and morality? Are religious attitudes direct adaptations for human social life or evolutionary by-products of other predispositions? Is the moral sense an independent capacity, or do we need supernatural beliefs in order to be moral?

In his psychological studies of the effect of religion on altruistic behavior, Ara Norenzayan addresses three questions. The first is whether religious attitudes shape and stimulate prosociality and moral behavior. The answer is yes, probably as a result of the awareness of an omniscient supernatural power. Secondly, do all religions concern themselves with moral behavior? This is indeed true for all organized religions in large-scale societies, but the results are different for small-scale human groups. Norenzayan then asks as third question whether religion is necessary for morality? He argues that the answer is no, because, amongst other reasons, moral understanding emerges during child development well before religion is acquired, and citizens of secularized societies fail to share the distrust of many believers that atheists cannot be moral.

Vittorio Girotto, Telmo Pievani and Giorgio Vallortigara — crossing their different backgrounds in psychology, evolution, and cognitive ethology — discuss the limits of a functional view of supernatural beliefs according to which these beliefs are adaptations shaped by natural selection for human social life, favoring cooperation and group cohesion. They compare this hypothesis to the alternative that supernatural thinking could be a secondary effect of powerful cognitive predispositions originally shaped for different reasons in previous ecological and social niches. Humans may be predisposed to detect causes, agents, and designs where there are none. The authors discuss the role of such animistic and teleological predispositions and sociocultural factors in promoting the diffusion of supernatural beliefs, stressing the process of functional cooptation of phylogenetically conserved traits in new social and cultural niches.

The Editors

[When citing this chapter, refer to Behaviour 151 (2014) 365–384]

Does religion make people moral?

Ara Norenzayan *

Department of Psychology, University of British Columbia,
2136 West Mall, Vancouver, BC, Canada V6T 1Z4
*Author's e-mail address: ara@psych.ubc.ca

Accepted 15 September 2013; published online 5 December 2013

Abstract

I address three common empirical questions about the connection between religion and morality: (1) Do religious beliefs and practices shape moral behavior? (2) Do all religions universally concern themselves with moral behavior? (3) Is religion necessary for morality? I draw on recent empirical research on religious prosociality to reach several conclusions. First, awareness of supernatural monitoring and other mechanisms found in religions encourage prosociality towards strangers, and in that regard, religions have come to influence moral behavior. Second, religion's connection with morality is culturally variable; this link is weak or absent in small-scale groups, and solidifies as group size and societal complexity increase over time and across societies. Third, moral sentiments that encourage prosociality evolved independently of religion, and secular institutions can serve social monitoring functions; therefore religion is not necessary for morality. Supernatural monitoring and related cultural practices build social solidarity and extend moral concern to strangers as a result of a cultural evolutionary process.

Keywords

religion, morality, culture, evolution, cooperation.

1. Introduction

Religion and morality are popular, complex and intensely controversial topics. So the intersection of the two is a hotly debated issue. Arguments about what, if anything, religion has to do with morality, have been raging for a long time. The idea that religions facilitate acts that benefit others at a personal cost has a long intellectual history in the social sciences (e.g., Darwin, 1859/1860; Durkheim, 1915/1995) and is a central idea in debates about the evolutionary origins of religions (Wilson, 2002; Sosis & Alcorta, 2003; Norenzayan & Shariff, 2008; Atran & Henrich, 2010; Bering, 2011).

However, this idea remains controversial, and has been critiqued by both opponents of religion (e.g., Dawkins, 2006; Dennett, 2006), as well as by behavioral scientists interested in the roots of morality as well as religion (e.g., Baumard & Boyer, 2013; de Waal, 2013).

There are several key empirical claims underlying this debate that are being actively investigated in the fast-moving evolutionary studies of religion. In this brief article, I bring together findings from experimental social psychology, cultural anthropology, behavioral economics, and history, and address three related but distinct questions about religion and morality that are at the core of this debate. These three questions are: (1) do religious beliefs and practices have any causal impact on moral behavior? (2) Do all religions universally prescribe moral behavior? (3) Is religion necessary for morality?

I examine these three questions in light of the empirical evidence. In doing so, I present a theory that explains the connection between religion and prosocial behavior (a key aspect of morality) as the outcome of an autocatalytic historical process that is shaped by cultural evolution — non-genetic, socially transmitted changes in beliefs and behaviors. I start with a brief summary of this argument. The specific details, as well as the wide-ranging evidence that this argument rests on, can be found elsewhere (Norenzayan & Shariff, 2008; Atran & Henrich, 2010; Norenzayan, 2013; Slingerland et al., in press). Then I outline the implications of this argument for the above three questions, while being mindful that other related, but distinct perspectives on the evolutionary origins of religion may have different takes on the religion and morality debate (e.g., Bering, 2011; Bulbulia & Sosis, 2011; Schloss & Murray, 2011; Bloom, 2012; Baumard & Boyer, 2013).

The starting point is that religious beliefs and practices emerged as cognitive side-effects of a set of biases rooted in mental architecture, such as the intuition that minds can operate separate from bodies (mind-body dualism), and that people and events exist for a purpose (teleology). Once intuitions about supernatural beings and ritual-behavior complexes were in place, rapid cultural evolution facilitated a process of coevolution between societal size and complexity on one hand, and devotional practices to Big Gods on the other — increasingly powerful, interventionist, and morally concerned supernatural monitors of the expanding group who demand unwavering commitment and loyalty. Over historical time in the last ten-to-twelve millenia, this led to — in some places but not others — the gradual linking up of religious beliefs and practices with prosocial tendencies, or religious prosociality (Norenzayan & Shariff, 2008; Norenzayan, 2013). In turn, belief in these moralizing deities and related social commitment devices cascaded around the world with these ever-expanding, culturally spreading groups.

In this way, religious prosociality helps explain the scientific puzzle of large-scale cooperation in humans. This is a puzzle for three reasons. Despite the fact that for most of their evolutionary history human beings lived in small bands of foragers (who had in turn descended from primate troops), today, the vast majority of humans live in large, anonymous, yet intensely cooperative societies (Seabright, 2004). Second, this change happened rapidly and very recently, that is, in the last 12000 years. Third, while human beings share with their primate relatives many cooperative instincts (de Waal, 2008), the scope and intensity of large-scale cooperation in humans are unknown in other species (Richerson & Boyd, 2005).

The central idea, then, is that the spread of prosocial religions in the last twelve millenia has been an important shaper of large-scale societies where anonymous interactions are essential to the social fabric. Importantly, it is not, and has not been, the only force leading to the scaling up of the cooperative sphere. Cultural norms for cooperation with strangers, as well institutions that enforce trust and cooperation, by for example, introducing third-party punishment (Herrmann ct al., 2008) also have broadened the moral sphere. However, institutions such as courts, police, and other contract-enforcing mechanisms are not always effective, have developed rather recently and only in some places. In the developing world, these institutions lack credibility, and therefore in the majority of the world, religion continues to thrive as an important source of cooperation and trust among strangers (e.g., Norris & Inglehart, 2004). But when they have succeeded, these institutions have replaced the community-building functions of prosocial religions. Effectively, these secular societies, guided by secular mechanisms for norm-enforcement, have climbed the ladder of religion and then kicked it away.

Religious prosociality binds unrelated strangers together, but, contrary to many theological teachings, there is little reason to expect that this prosociality is actually extended without limits to everyone. The same forces that cement and expand social solidarity within the group also have the potential to feed the flames of intolerance and conflict between rival religious communities, particularly when one's group is seen to be under threat by these groups or by nonbelievers. The precise boundaries of religious prosociality, and its role in fueling conflict, are important open questions for scientific study. But the seeming paradox that religion is both the handmaiden of cooperation within groups, and conflict and prejudice between groups, can be explained by the same psychological mechanisms that religions exploit to create social solidarity (Norenzayan, 2013; see also Haidt, 2012; Bloom, 2012).

Before we begin, two further clarifications are in order about the two loaded terms that are at the center of this debate: 'religion' and 'morality'. Let's begin with 'religion' first. The theoretical argument I offer here about religion combines two powerful ideas: first, that the intuitions that underlie religious beliefs and practices, such as commitment to supernatural beings, the sacred, and ritual behaviors, are natural byproducts of built-in cognitive tendencies that are likely to have innate components (e.g., Boyer, 2001; Atran & Norenzayan, 2004; Barrett, 2004); second, that once religious intuitions or templates are in place and produce a constrained but diverse set of beliefs, their content undergoes rapid cultural evolution such that some cultural variants spread at the expense of others (Richerson & Boyd, 2005; Norenzayan, 2013).

Taken together, these two ideas sharpen the debate about what religion is and how it can be studied scientifically. In the humanities, there is a long tradition of debating (apparently without any clear resolution) the definition of the term 'religion' (see, for example, Clarke & Byrne, 1993; Stausberg, 2010). However, in the evolutionary perspective that motivates the argument presented here, and in agreement with much of the cognitive science approach to religion, it becomes clear that 'religion' is not a natural kind category or a definable concept, therefore semantic debates about how to define religion are not scientifically productive. Rather, the term 'religion' is more accurately seen as a convenient label, pointing towards a package of (precisely operationalized) beliefs and behaviors. This package is assembled over historical time, taking on different shapes in different cultural and historical contexts. From a cultural evolutionary perspective, then, the scientific project of explaining religion is not only to account for the universal features of religion found in every human society, but to also explain the often dramatic cultural changes that we see in the 'religious package' found in the historical and ethnographic record (e.g., Swanson, 1964; Roes & Raymond, 2003).

Similar to 'religion', 'morality' is also a hotly debated concept, and there are many important and unresolved issues (Doris et al., 2012). However, once again, for the purposes of the discussion here, we need not agree on the clear demarcation (necessary and sufficient conditions) of what constitutes morality. Even if such conditions existed and were similar across cultures — an important but separate issue — we can proceed by being precise about the components of beliefs and behaviors that are under investigation and that fall under the rubric of morality. Taking into account these considerations, the evolutionary perspective presented here sees human moral psychology — as well as religion — as a natural phenomenon that is the converging product of genetic and cultural inheritance. At the broadest level, then, morality can be conceptualized as "…interlocking sets of values, virtues, norms, practices, identities, institutions, technologies, and evolved psychological mechanisms that work together to suppress or regulate self-interest and make cooperative societies possible" (Haidt, 2012, p. 270). From an evolutionary standpoint, morality is therefore intimately linked to the problem of how large, anonymous, but cooperative societies solve the problem of free riding.

2. Question 1: Do religious beliefs and practices encourage moral behavior?

Does religion encourage prosocial behavior? Here I discuss and highlight evidence drawn from three different social science literatures based on different methods that address this question. As is the case for any empirical science on an important question, the conclusions from each of these literatures has its limitations, and is best considered in combination with other evidence using other approaches to reach firm conclusions.

One traditional approach to answer this question is based on sociological surveys. American survey respondents who frequently pray and attend religious services (regardless of religious denomination) reliably report more prosocial behavior, such as more charitable donations and volunteerism (Brooks, 2006). Brooks reports, for example, that in the United States, 91% of people who attend religious services weekly or more often report donating money to charities, compared to only 66% those who attend religious services a few times a year or less. However, surveys, as useful as they are, suffer from methodological limitations and are open to alternative interpretations (for a critique, see Norenzayan & Shariff, 2008). One serious limitation, for example, is that people often exaggerate socially desirable behaviors (such as how much they volunteer or give to charity). This is particularly an issue here since religiosity itself increases social desirability concerns (Gervais & Norenzayan, 2012a). Therefore, the gap found in these surveys between believers and non-believers may not reflect 'doing good' as much as it may reflect 'appearing good'.

A second approach has assessed whether self-reports of religiosity predict actual prosocial behavior measured under controlled conditions. These studies have reported mixed findings. Some studies have found no associations between religious involvement and prosocial tendencies; others have found that religious involvement does predict more prosocial behavior, but only when the prosocial act could promote a positive image for the participant, either in their own eyes or in the eyes of observers (Batson et al., 1993). Other studies, conducted outside of North America and Europe, have found a reliable association between intensity of religious participation or involvement, and willingness to cooperate or contribute to a common pool (e.g., Sosis & Ruffle, 2003; Henrich et al., 2010; Soler, 2012).

A third approach has gone beyond survey and correlational methods and has taken advantage of combining two techniques; one, cognitive priming from experimental psychology to activate religious thoughts, and two, games from behavioral economics, where actual prosocial behavior with monetary incentives can be measured in controlled conditions. If religious thinking has a causal effect on prosocial tendencies, then experimentally-induced religious reminders should increase prosocial behavior in controlled conditions. If so, subtle religious reminders may reduce cheating, curb selfish behavior, and increase generosity towards strangers. This hypothesis is gaining increasing support (for a summary, see Norenzayan et al., in press; see also Norenzayan & Shariff, 2008). In one experiment (Shariff & Norenzayan, 2007; see Figure 1), adult non-student participants were randomly assigned to three groups: participants in the religious prime group unscrambled sentences that contained words such as God, divine, and spirit; the secular prime group unscrambled sentences with words such as civic, jury, police; and the control group unscrambled sentences with entirely neutral content. Each participant subsequently played an anonymous double-blind one-shot Dictator Game. (Post-experimental debriefing showed that participants showed no awareness of the priming concepts, or awareness of the hypothesis of the study.) Compared to the control group, nearly twice as much money

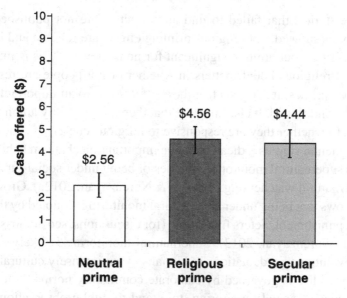

Figure 1. Priming religious concepts increased offers in the Dictator Game among Canadian adults; priming secular concepts had a comparable effect. The results showed not only a quantitative increase in generosity, but also a qualitative shift in giving norms. In the control group, the modal response was selfishness, a plurality of players pocketed all $10. In the religious and secular priming conditions, the mode shifted to fairness, a plurality of players split the money evenly ($N = 75$). Figure from Norenzayan & Shariff (2008).

was offered by subjects in the religious prime group. Of particular interest, the secular prime group showed the same pattern as the religious prime group, suggesting that secular mechanisms, when they are available, can also encourage generosity. Religious primes also reduce cheating among students in North America (Randolph-Seng & Nielsen, 2007), as well as in children (Piazza et al., 2011). McKay et al. (2011) found that subliminal religious priming increased third-party costly punishment of unfair behavior in a Swiss religious sample (see also Laurin et al., 2012). In these studies, individual differences in religious involvement or belief were unrelated to prosocial behavior.

Pooling all known studies together, a recent meta-analysis was conducted to assess the extent to which these effects are replicable (A.F. Shariff, A. Willard, T. Andersen & A. Norenzayan, data not shown). Overall, religious priming was found to increase prosocial behavior, with a moderate average effect size. The effect remained robust after estimating and accounting for the file-drawer effect or publication bias in psychology (that is, the possibility that some studies that failed to find any effects were not published). Further analyses showed that religious priming effects are reliable and large for strong believers, but are non-significant for nonbelievers. This is important, because, if religious belief matters in whether or not people are responsive to religious primes, it suggests that these effects are, to an important degree, culturally conditioned. It also suggests that there is variability among nonbelievers as to whether they are responsive to religious cues.

Experimental studies indicate that one important mechanism behind these effects is supernatural monitoring, or cues of being under social surveillance by a supernatural watcher (e.g., Gervais & Norenzayan, 2012a). Growing evidence shows that being under supernatural monitoring, coupled by the threat of divine punishment, deters free-riding (for discussions, see Schloss & Murray, 2011; Norenzayan, 2013). Supernatural monitoring is likely rooted in ancient evolutionary adaptations in humans — an intensely cultural species whose social life is governed by elaborate community norms — to be sensitive to cues of social monitoring, to attend to public observation, and to anticipate punishment for norm-violations (Henrich & Henrich, 2007). As the saying goes, 'watched people are nice people'. A wide range of laboratory and field studies shows that social surveillance, or the expectation of monitoring and accountability increases prosocial tendencies (see, for example, Fehr & Fischbacher, 2003; Bateson et al., 2006).

Therefore, even when no one is watching, believers are more likely to act nicely towards strangers when they feel that a Big God is watching. It is also likely that there are additional independent mechanisms underlying religious prosociality that converge with supernatural monitoring. Other candidate mechanisms that are being investigated include participation in intense rituals (Xygalatas et al., 2013), and synchronous movement and music (Wiltermuth & Heath, 2009: but see Cohen et al., in press).

Importantly for debates about religion and morality, these studies show that when religious reminders are absent, believers and non-believers — especially those from societies with strong rule of law — are equally prosocial towards strangers. Other studies that rely on situational priming bolster this conclusion. Xygalatas (2013) randomly assigned Hindu participants in Mauritius to play a common pool resource game either in a religious setting (a temple) or in a secular setting (a restaurant). Participants preserved the shared pool of money more when they played the game in the temple compared to when they played in the restaurant. Individual differences in the intensity of religiosity were unrelated to sharing. Malhotra (2008) took advantage of the fact that for Christians, reminders of religion are more salient on Sundays than on other days of the week. He measured responsiveness to an online charity drive over a period of several weeks. Christians and nonbelievers were equally likely to give to charity except on Sundays, when Christians were three times more likely to give. These results suggest that religious prosociality is context-specific; in other words, the 'religious situation' is stronger than the 'religious disposition'.

In summary, this experimental literature, complements other evidence, allowing researchers to test a set of more specific hypotheses about religious prosociality, and with more experimental rigor that allows for causal inference. The evidence increasingly shows that there is an arrow of causality that goes from religion to a variety of prosocial behaviors, including generosity, honesty, cooperation, and altruistic punishment (Norenzayan & Shariff, 2008; Norenzayan et al., in press). Despite these important insights, the experimental priming literature is limited in that it is mostly based on samples from Western industrialized societies. These studies limit inferences about the cultural generalizability of these effects, an issue that is addressed next.

3. Question 2: Are all religions about morality?

It is believed by many that supernatural agents specifically, and all religions more broadly, are inherently about morality — that all religions concern themselves with regulating moral affairs within a community. Particularly for those immersed in the Abrahamic traditions — believers and nonbelievers alike — there is a powerful intuitive appeal to this idea. After all, in these cultures, as well as in other world religions such as Buddhism and Hinduism, religion *is* intensely about regulating human morality. However, the ethnographic and historical record contradicts the claim that this linkage is a cultural universal. One of the early pioneers of the comparative study of religion, Guy Swanson (1964: 153) concluded, "The people of modern Western nations are so steeped in these beliefs which bind religion and morality, that they find it hard to conceive of societies which separate the two. Yet most anthropologists see such a separation as prevailing in primitive societies."

In small-scale societies, people must tackle an extensive variety of cooperative challenges, and therefore they are guided by a sophisticated set of local moral norms that apply to a wide range of domains, including food sharing, caring of offspring, kinship relations, marriage, leveling of risk, and mutual defense. Moreover, these groups vary in important ways, such as in population size and density, technology, and sedentary lifestyle (Kelly, 1995; Powell et al., 2009). While recognizing these important complexities, ethnographic observations support Swanson's claim — they show that in these small-scale societies, religion's moral scope, if any, is minimal; the gods tend to have limited omniscience and limited moral concern; they may want rituals and sacrifices, but care little about how people treat each other (Swanson, 1964; Boyer, 2001; Marlowe, 2010; Purzycki, 2011). Purzycki (2011), for example, reports that for pastoralists in Tuva culture in Siberia, local 'spirit masters' known as *Cher eezi*, are pleased by ritual offerings, and are angered by over-exploitation of resources, but only the ones that they directly oversee. They exert their powers in designated areas found in ritual cairns known as *ovaa*. *Cher eezi* do not see far and cannot intervene in distant places. While the *Cher eezi* have some powers and care about some things, in foraging societies, typically the gods are even more distant and indifferent. Anthropologist Frank Marlowe (2010), who has lived with Hadza foragers of Tanzania, describes Hadza religion this way:

"I think one can say that the Hadza do have a religion, certainly a cosmology anyway, but it bears little resemblance to what most of us in complex societies (with Christianity, Islam, Hinduism, etc.) think of as religion. There are no churches, preachers, leaders, or religious guardians, no idols or images of gods, no regular organized meetings, no religious morality, no belief in an afterlife — theirs is nothing like the major religions."

These observations are important — if religious prosociality was a pan-human genetic adaptation, it should be found everywhere, especially among foraging societies that give us the best (though imperfect) clues we have of ancestral human conditions. But the ethnographic record further shows that, although all known societies have gods and spirits, there is a cultural gradient in the degree to which they are (1) omniscient, (2) interventionist and (3) morally concerned. Stark (2001), for example, found that less than a quarter of the cultures in one ethnographic database of the world's cultures have a Big God who is involved in human affairs and cares about human morality. But this cultural variability is non-random; it covaries systematically with societal size and complexity (Roes & Raymond, 2003; Johnson, 2005). As group size increases, the odds increase of the existence of one or several Big Gods — omniscient, all-powerful, morally concerned deities who directly regulate moral behavior and dole out punishments and rewards.

In the cultural evolutionary perspective, these observations make sense. In small-scale societies, where face-to-face interactions are typical, people build cooperative communities that draw on kin altruism, reciprocity, and a rich repertoire of local cultural norms — without needing to lean on watchful gods. But as societies scale up and groups get too large, anonymity rapidly invades interactions; otherwise cooperative human behaviors begin to erode (e.g., Henrich & Henrich, 2007). It is precisely in these anonymous societies that, widespread belief in watchful gods, if adopted, could contribute to maintaining large-scale cooperation. The direct implication of this fact, which begs for scientific explanation, is that despite starting off as a rare cultural variant, belief in these Big Gods spread so successfully that the vast majority of the world's believers belong to religions with such gods (Norenzayan, 2013). The linking up of religion with morality, then, is a cultural development that emerged over historical time in some places. Cultural variants of gods that contributed to the creation and expansion of social solidarity were more likely to proliferate. Although aspects of both what we call 'religion' and 'morality' have innate components (see below), the linking of the two appears to be a cultural shift, not a fixed part of humanity's genetic inheritance.

There is further cross cultural, archeological, and historical evidence supporting this cultural evolutionary hypothesis of religious prosociality driven by passionate devotion to Big Gods. In a wide-ranging investigation spanning 15 societies of pastoralists and horticulturalists, Henrich et al. (2010) specifically tested the idea that participation in prosocial religions with Big Gods encourages more prosocial behavior compared to participation in local religions that typically do not have gods with wide moral scope. Henrich and colleagues found that, controlling for age, sex, household size, community size, and a wide range of other socio-demographic variables, endorsement of religions with Big Gods increased offers in the Dictator Game by 6%, and in the Ultimatum game by 10% (given a standardized stake equaling 100). These are substantial effects, once we realize that (1) the range of behavioral variation in these games is quite restricted (in both games, rarely people offer more than 50% of the stake); and (2) other known contributing factors to prosocial behavior were accounted for (therefore, these effect sizes are specific to religion). The other key finding was that greater market integration, that is, experience with economic exchange with strangers, also led to greater prosocial behavior. Once again, prosocial religions are an important factor, but not the only factor, in encouraging prosociality with strangers.

Available archeological evidence, though limited, is consistent with these cross-cultural findings. Although devotion to Big Gods does not typically reveal material traces before writing emerged, the archeological record contains several hints that related practices, such as the expansion of regular rituals and the construction of religiously significant monumental architecture co-occurred as populations exploded, political complexity increased, and agriculture spread (Cauvin, 1999; Marcus & Flannery, 2004). Evidence for this can be found in Çatalhöyük, a 9500-year-old Neolithic site in southern Anatolia (e.g., Whitehouse & Hodder, 2010). The on-going excavation of Göbekli Tepe, a 11 000-year-old site of monumental architecture with religious significance, suggests that it may have been one of the world's first temples, where hunter-gatherers possibly congregated and engaged in organized religious rituals (Schmidt, 2010).

One of the best-documented historical case studies looks at the Big God of the Abrahamic traditions. Textual evidence reveals the gradual evolution of the Abrahamic God from a tribal war god with limited social and moral concern, to the unitary, supreme, moralizing deity of Judaism, and two of the world's largest religious communities — Christianity and Islam (a summary of this evidence can be found in Wright, 2009). Another relevant case study is Chinese civilization. There is an active debate about the precise role of supernatural monitoring and other secular mechanisms in the moral order of the emerging and evolving Chinese polity (Clark & Winslett, 2011; Paper, 2012). Nevertheless, even there, evidence from early China shows that supernatural monitoring and punishment played an increasingly important role in the emergence of the first large-scale societies in China (see Clark & Winslett, 2011; Slingerland et al., in press). In summary, there are important open questions and debates regarding the role of religious prosociality and other mechanisms in the ethnographic and historical record of the scaling up of societies over time. But these debates revolve around a statistical pattern that suggests that, religion's role in regulating moral affairs in large societies has been a cultural process that coalesced over time, primarily where anonymous societies took shape and expanded.

4. Question 3: Is religion necessary for morality?

While there is mounting evidence that reminders of supernatural monitors and related practices encourage prosociality, another idea, that without religion, there could be no morality, also deserves careful attention because it is as widely believed as it is mistaken. This is not just the personal opinion of Dr. Laura Schlessinger (an influential public media personality in America) who infamously claimed, "it's impossible for people to be moral without a belief in God. The fear of God is what keeps people on the straight and narrow" (Blumner, 2011). It is 'common wisdom' among many religious believers, and is a primary reason why distrust of atheism is rampant among them (for evidence and reviews, see Gervais et al., 2011; Gervais & Norenzayan, 2013). It also appears to be one of the key reasons why believers would rather trust people who believe in the 'wrong god' (that is, someone of another religion), than they would trust people of their own culture who believe in no god, that is, atheists (Edgell et al., 2006; Norenzayan, 2013).

Even a major Enlightenment figure as John Locke shared this intuition. In the landmark *Letter Concerning Toleration* (1983/1689), a foundational document that ushered the idea of religious tolerance of minority groups, Locke defended religious diversity, and then excluded atheists from moral protection:

> "Those are not at all to be tolerated who deny the being of a God. Promises, covenants, and oaths, which are the bonds of human society, can have no hold upon an atheist. The taking away of God, though but even in thought, dissolves all."

Despite its widespread appeal, this view does not fit the facts at least for two important reasons. First, core human moral instincts, such as empathy, compassion, and shame are much more ancient than religiously motivated prosociality, and are deeply rooted in the primate heritage (de Waal, 2013), and some of the precursors of these instincts can be found even in the mammalian brain (Churchland, 2012). Some precursors of moral instincts, such as capacities for emotional contagion, consolation, and grief have been found in chimpanzees as well as other species, such as elephants (de Waal, 2008).

These early building blocks of moral psychology draw on innate instincts and emotions rooted in evolutionary adaptations, such as kinship psychology and the caring of offspring (empathy, compassion), reciprocity (guilt, anger), and deference towards dominance hierarchies (shame, pride). Everyone, believers and disbelievers alike, have them. In one study that looked at whether feelings of compassion led to prosocial behavior among believers and non-believers, Saslow et al. (2012) found that (1) religiosity and feelings of compassion were statistically unrelated; and (2) for nonbelievers, the greater the feelings of compassion were, the more prosocial their behavior was; (3) however, among believers, feelings of compassion were unrelated to prosocial behavior. Although more studies are needed to reach firm conclusions, these results suggest that if anything, compassion may be more strongly linked to prosociality among non-believers. While we have ample evidence that supernatural monitoring provided by world religions encourage prosociality, these preliminary data by Saslow et al. show that, if anything, moral emotions such as compassion are more strongly linked up with prosociality in non-believers, which could explain why, typically believers and nonbelievers do not differ in prosociality unless religious reminders are present in the situation.

There is additional evidence that suggests that moral intuitions are primary and likely have innate components, as even preverbal babies have the early precursors of these intuitions. For example, by 6-months of age, babies show a preference for an individual who helps and an aversion to an individual who obstructs someone else's goal (Hamlin et al., 2007). Eight-month old babies not only prefer prosocial individuals, but they also prefer individuals who act harshly towards an antisocial individual (Hamlin et al., 2011). However, these moral emotions are intuitively directed towards family members and close friends, therefore, socializing children and adults to extend them more broadly is possible, but is not a given; it is facilitated by cultural norms about how to treat others. In support of this idea, in economic game studies with adult participants in large-scale societies, some amount of prosocial tendencies remain even when experimenters go at great lengths to ensure anonymity and lack of accountability (Henrich & Henrich, 2006). In this way, societies develop norms for kindness, fairness, and other virtues that harness the moral emotions to expand the moral circle (see, for example, Kitcher, 2011; Singer, 2011).

What about prosocial behavior among strangers that is motivated by social monitoring incentives? Here too, the view that religion is necessary for morality is mistaken, because it overlooks the fact that supernatural monitoring and punishment are not the only game in town — in societies with strong rule of law, there are other social monitoring incentives and institutions that encourage prosociality. In particular, recent psychological and sociological evidence show that people exposed to strong secular rule of law are more trusting and prosocial than people exposed to weak or non-existent rule of law (Kauffman et al., 2003; Herrmann et al., 2008).

Where there are strong institutions that govern public life, that is, where people are reassured that, contracts are enforced, competition is fair, and cheaters will be detected and punished, there are high levels of trust and co-operation among strangers. Interestingly, the role of religion in public life declines as societies develop secular alternatives that serve similar functions (Norris & Inglehart, 2004; Norenzayan & Gervais, 2013). This means that atheists, as well as theists who are socialized in such secular societies are prosocial without (or in addition to) immediately being motivated by religion. This also explains why Scandinavian societies are some of the world's least religious societies but also the most cooperative and trusting ones (Zuckerman, 2008).

Secular sources of prosociality not only dampen religious zeal; they also appear to weaken believers' intuition that religion is necessary for morality. Thus, religious distrust of atheists, although common among many believers, is not immutable. All else being equal, believers who live in countries with strong secular institutions (as measured by the World Bank's rule of law index) are more willing to trust atheist politicians than equally devoted believers who live in countries with weak institutions (Norenzayan & Gervais, 2013). These cross cultural survey findings are also supported by experimental evidence, where causal pathways can be identified with more confidence. In studies done in Canada and the USA (countries that have strong rule of law), experimentally induced reminders of concepts such as court, police, and judge, that previously were found to increase generosity (Shariff & Norenzayan, 2007), also reduced believers' distrust of atheists, presumably by undermining the intuition that religion is necessary for morality, and by highlighting the fact that there are other, secular incentives that motivate prosocial behavior (Gervais & Norenzayan, 2012b).

5. Coda

So does religion make people moral? This is a complex question with a complex answer. If, by 'moral', we mean 'prosocial towards strangers of one's

imagined moral community', the growing evidence suggests that supernatural monitoring and related practices indeed are one factor that makes people
act more prosocially towards others. However, this prosociality has its limits,
it can turn toxic when religious groups feel threatened by rival groups, and
in believers' distrust and exclusion of non-believers (Gervais & Norenzayan, 2013). Moreover, not all cultural variants of religion make people moral
in this sense, and importantly, the best evidence we have suggests that the
origins of religious cognition are unrelated to the origins of morality. Religious prosociality is therefore best explained as a cultural process, where
supernatural beings, over time and in some places, became more omniscient,
more omnipotent, and more moralizing. In doing so these gods spread by
galvanizing large-scale cooperation at an unprecedented scale.

Importantly, these facts about religious prosociality are not incompatible
with non-religious sources of moral systems, that is, secular humanism. This
is partly because human beings are endowed with natural moral instincts
that, although intuitively are directed towards family, friends, and allies, can,
under the right conditions, be harnessed by cultural evolution to broaden
the moral scope to include strangers. Moreover, supernatural monitoring
draws on pre-existing social monitoring mechanisms that promote large-
scale cooperation once secular societies develop institutions that are capable
of extending and solidifying the rule of law. Secular societies with effective
institutions promote strong cooperative norms, and this is precisely where
the vast majority of atheists live. Moreover, these institutions have replaced
religious sources (and in some cases such as Northern Europe, much more
effectively), and given birth to secular humanism, or a set of norms grounded
in morality without reliance on gods. In some cases, majority atheist societies
have become the most cooperative, peaceful, and prosperous societies in
history (Zuckerman, 2008).

Finally, prosocial religions have been important cultural solutions that
contributed to the creation of anonymous, moral communities, but clearly
they are not necessary for morality. The same forces of cultural evolution
that gave rise to prosocial religions with Big Gods also have, more recently,
given rise to secular mechanisms that promote large-scale cooperation and
trust. These social monitoring and norm-enforcement mechanisms, coupled
with an innately given repertoire of moral emotions that can be harnessed to
widen the scope of moral concern, have fashioned a new social phenomenon,
perhaps even a novel social transition in human history: cooperative moral
communities without belief in Big Gods.

References

Atran, S. & Henrich, J. (2010). The evolution of religion: how cognitive by-products, adaptive learning heuristics, ritual displays, and group competition generate deep commitments to prosocial religions. — Biol. Theor. 5: 18-30.

Atran, S. & Norenzayan, A. (2004). Religion's evolutionary landscape: counterintuition, commitment, compassion, communion. — Behav. Brain Sci. 27: 713-770.

Barrett, J.L. (2004). Why would anyone believe in God? — AltaMira Press, Walnut Creek, CA.

Bateson, M., Nettle, D. & Roberts, G. (2006). Cues of being watched enhance cooperation in a real-world setting. — Biol. Lett. 2: 412-414.

Batson, C.D., Schoenrade, P. & Ventis, W.L. (1993). Religion and the individual: a social-psychological perspective. — Oxford University Press, New York, NY.

Baumard, N. & Boyer, P. (2013). Explaining moral religions. — Trends Cogn. Sci. 17: 272-280.

Bering, J. (2011). The belief instinct: the psychology of souls, destiny, and the meaning of life. — W.W. Norton, New York, NY.

Bloom, P. (2012). Religion, morality, evolution. — Ann. Rev. Psychol. 63: 179-199.

Blumner, R.E. (2011). Goodness without God. — St. Petersburg Times, 8 July 2011.

Boyer, P. (2001). Religion explained. — Basic Books, New York, NY.

Brooks, A.C. (2006). Who really cares: the surprising truth about compassionate conservatism. — Basic Books, New York, NY.

Bulbulia, J. & Sosis, R. (2011). Signalling theory and the evolutionary study of religions. — Religion 41: 363-388.

Cauvin, J. (1999). The birth of the gods and the origins of agriculture. Trans. T. Watkins. — Cambridge University Press, Cambridge.

Churchland, P.S. (2012). Braintrust: what neuroscience tells us about morality. — Princeton University Press, Princeton, NJ.

Clark, K.J. & Winslett, J.T. (2011). The evolutionary psychology of Chinese religion: Pre-Qin high gods as punishers and rewarders. — J. Am. Acad. Relig. 79: 928-960.

Clarke, P. & Byrne, P. (1993). Religion defined and explained. — Macmillan Press, London.

Cohen, E., Mundry, R. & Kirschner, S. (in press). Religion, synchrony, and cooperation. — Relig. Brain Behav.

Darwin, C. (1859/1860). On the origins of species by means of natural selection. — John Murray, London.

Dawkins, R. (2006). The God delusion. — Houghton Mifflin, Boston, MA.

Dennett, D.C. (2006). Breaking the spell. — Viking, New York, NY.

de Waal, F. (2008). Putting the altruism back into altrusim: the evolution of empathy. — Annu. Rev. Psychol. 59: 279-300.

de Waal, F. (2013). The bonobo and the atheist: in search of humanism among the primates. — W.W. Norton, New York, NY.

Doris, J., Harman, G., Nichols, S., Prinz, J., Sinnott-Armstrong, W. & Stich, S. (2012). The moral psychology handbook. — Oxford University Press, Oxford.

Durkheim, E. (1915/1995). The elementary forms of the religious life. — Free Press, New York, NY.

Edgell, P., Gerteis, J. & Hartmann, D. (2006). Atheists as "other": moral boundaries and cultural membership in American society. — Am. Sociol. Rev. 71: 211-234.

Fehr, E. & Fischbacher, U. (2003). The nature of human altruism. — Nature 425: 785-791.

Gervais, W.M. & Norenzayan, A. (2012a). Like a camera in the sky? Thinking about God increases public self-awareness and socially desirable responding. — J. Exp. Soc. Pychol. 48: 298-302.

Gervais, W.M. & Norenzayan, A. (2012b). Reminders of secular authority reduce believers' distrust of atheists. — Psychol. Sci. 23: 483-491.

Gervais, W.M. & Norenzayan, A. (2013). Religion and the origins of anti-atheist prejudice. — In: Intolerance and conflict: a scientific and conceptual investigation (Clarke, S., Powell, R. & Savulescu, J., eds). Oxford University Press, Oxford, p. 126-145.

Gervais, W.M., Shariff, A.F. & Norenzayan, A. (2011). Do you believe in atheists? Distrust is central to anti-atheist prejudice. — J. Personality Soc. Psychol. 101: 1189-1206.

Haidt, J. (2012). The righteous mind: why good people are divided by politics and religion. — Pantheon Books, New York, NY.

Hamlin, J.K., Wynn, K. & Bloom, P. (2007). Social evaluation by preverbal infants. — Nature 450: 557-559.

Hamlin, J.K., Wynn, K., Bloom, P. & Mahajan, N. (2011). How infants and toddlers react to antisocial others. — Proc. Natl. Acad, Sci. USA 108: 19931-19936.

Henrich, J., Ensimger, J., McElreath, R., Barr, A., Barrett, C., Bolyanatz, A., Cardenas, J.C., Gurven, M., Gwako, E., Henrich, N., Lesorogol, C., Marlowe, F., Tracer, D. & Ziker, J. (2010). Markets, religion, community size, and the evolution of fairness and punishment. — Science 327: 1480-1484.

Henrich, N.S. & Henrich, J. (2007). Why humans cooperate: a cultural and evolutionary explanation. — Oxford University Press, Oxford.

Herrmann, B., Thöni, C. & Gächter, S. (2008). Antisocial punishment across societies. — Science 319: 1362-1367.

Johnson, D.D. (2005). God's punishment and public goods. — Human Nature 16: 410-446.

Kauffman, D.A., Kraay, A. & Mastruzzi, M. (2003). Governance matters III: Governance indicators for 1996–2002. — World Bank Econ. Rev. 18: 253-287.

Kelly, R.L. (1995). The foraging spectrum: diversity in hunter-gatherer lifeways. — Smithsonian Institution Press, Washington, DC.

Kitcher, P. (2011). The ethical project. — Harvard University Press, Cambridge, MA.

Laurin, K., Shariff, A.F., Henrich, J. & Kay, A.C. (2012). Outsourcing punishment to god: beliefs in divine control reduce earthly punishment. — Proc. Roy. Soc. Lond. B: Biol. Sci. 279: 3272-3281.

Locke, J. (1983). A letter concerning toleration. — Hackett, Indianapolis, IN.

Malhotra, D. (2008). (When) Are religious people nicer? Religious salience and the "Sunday Effect" on prosocial behavior. — Judgm. Decis. Making 5: 138-143.

Marcus, J. & Flannery, K.V. (2004). The coevolution of ritual and society: new 14C dates from ancient Mexico. — Proc. Natl. Acad, Sci. USA 101: 18257-18261.

Marlowe, F.W. (2010). The Hadza: hunter–gatherers of Tanzania. — University of California Press, Berkeley, CA.

McKay, R., Efferson, C., Whitehouse, H. & Fehr, E. (2011). Wrath of God: religious primes and punishment. — Proc. Roy. Soc. Lond. B: Biol. Sci. 278: 1858-1863.

Norenzayan, A. (2013). Big Gods: how religion transformed cooperation and conflict. — Princeton University Press, Princeton, NJ.

Norenzayan, A. & Gervais, W.M. (2013). The origins of religious disbelief. — Trends Cogn. Sci. 17: 20-25.

Norenzayan, A. & Shariff, A.F. (2008). The origin and evolution of religious prosociality. — Science 322: 58-62.

Norenzayan, A., Henrich, J. & Slingerland, E. (in press). Religious prosociality: a synthesis. — In: Cultural evolution (Richerson, P. & Christiansen, M., eds). MIT Press, Cambridge, MA.

Norris, P. & Inglehart, R. (2004). Sacred and secular: religion and politics worldwide. — Cambridge University Press, Cambridge.

Paper, J. (2012). Response to Kelly James Clark & Justin T. Winslett, "The evolutionary psychology of Chinese religion: Pre-Qin high gods as punishers and rewarders". — J. Am. Acad. Relig. 80: 518-521.

Piazza, J., Bering, J.M. & Ingram, G. (2011). "Princess Alice is watching you": children's belief in an invisible person inhibits cheating. — J. Exp. Child Psychol. 109: 311-320.

Powell, A., Shennan, S. & Thomas, M.G. (2009). Late Pleistocene demography and the appearance of modern human behavior. — Science 324: 1298-1301.

Purzycki, B.G. (2011). Tyvan cher eezi and the sociological constraints of supernatural agents' minds. — Relig. Brain Behav. 1: 31-45.

Randolph-Seng, B. & Nielsen, M.E. (2007). Honesty: one effect of primed religious representations. — Int. J. Psychol. Relig. 17: 303-315.

Richerson, P.J. & Boyd, R. (2005). Not by genes alone: how culture transformed human evolution. — University of Chicago Press, Chicago, IL.

Roes, F.L. & Raymond, M. (2003). Belief in moralizing gods. — Evol. Hum. Behav. 24: 126-135.

Saslow, L.R., Willer, R., Feinberg, M., Piff, P.K., Clark, K., Keltner, D. & Saturn, S.R. (2012). My brother's keeper? Compassion predicts generosity more among less religious individuals. — Soc. Psychol. Personal. Sci. 4: 31-38.

Schloss, J.P. & Murray, M.J. (2011). Evolutionary accounts of belief in supernatural punishment: a critical review. — Relig. Brain Behav. 1: 46-99.

Schmidt, K. (2010). Göbekli Tepe — the Stone Age Sanctuaries. New results of ongoing excavations with a special focus on sculptures and high reliefs. — Doc. Praehistor. XXXVII: 239-256.

Seabright, P. (2004). In the company of strangers: a natural history of economic life. — Princeton University Press, Princeton, NJ.

Shariff, A.F. & Norenzayan, A. (2007). God is watching you: priming god concepts increases prosocial behavior in an anonymous economic game. — Psychol. Sci. 18: 803-809.

Singer, P. (2011). The expanding circle: ethics, evolution and moral progress. — Princeton University Press, Princeton, NJ.

Slingerland, E., Henrich, J. & Norenzayan, A. (in press). The evolution of prosocial religions. — In: Cultural evolution (Richerson, P. & Christiansen, M., eds). MIT Press, Cambridge, MA.

Soler, M. (2012). Costly signaling, ritual and cooperation: evidence from Candomblé, an Afro-Brazilian religion. — Evol. Hum. Behav. 33: 346-356.

Sosis, R. & Alcorta, C. (2003). Signaling, solidarity, and the sacred: the evolution of religious behavior. — Evol. Anthropol. 12: 264-274.

Sosis, R. & Ruffle, B.J. (2003). Religious ritual and cooperation: testing for a relationship on Israeli religious and secular kibbutzim. — Curr. Anthropol. 44: 713-722.

Stark, R. (2001). Gods, rituals, and the moral order. — J. Sci. Stud. Relig. 40: 619-636.

Stausberg, M. (2010). Prospects in theories of religion. — Methods Theor. Stud. Relig. 22: 223-238.

Swanson, G.E. (1964). The birth of the Gods. — University of Michigan Press, Ann Arbor, MI.

Whitehouse, H. & Hodder, I. (2010). Modes of religiosity at Çatalhöyük. — In: Religion in the emergence of civilization (Hodder, I., ed.). Cambridge University Press, Cambridge, p. 122-145.

Wilson, D.S. (2002). Darwin's cathedral. — University of Chicago Press, Chicago, IL.

Wiltermuth, S.S. & Heath, C. (2009). Synchrony and cooperation. — Psychol. Sci. 20: 1-5.

Wright, R. (2009). The evolution of God. — Little Brown, New York, NY.

Xygalatas, D. (2013). Effects of religious setting on cooperative behaviour. A case study from Mauritius. — Relig. Brain Behav. 3: 91-102.

Xygalatas, D., Mitkidis, P., Fischer, R., Reddish, P., Skewes, J., Geertz, A.W., Roepstorff, A. & Bulbulia, J. (2013). Extreme rituals promote prosociality. — Psychol. Sci. 24: 1602-1605.

Zuckerman, P. (2008). Society without God. — New York University Press, New York, NY.

[When citing this chapter, refer to Behaviour 151 (2014) 385–402]

Supernatural beliefs: Adaptations for social life or by-products of cognitive adaptations?

Vittorio Girotto [a], **Telmo Pievani** [b,*] **and Giorgio Vallortigara** [c]

[a] Department of Architecture and Arts, University IUAV of Venice, Santa Croce 191, Tolentini, 30135 Venice, Italy
[b] Department of Biology, University of Padua, Via U. Bassi 58/B, 35131 Padua, Italy
[c] Centre for Mind/Brain Sciences, University of Trento, Palazzo Fedrigotti, Corso Bettini 31, 38068 Rovereto Trento, Italy
*Corresponding author's e-mail address: dietelmo.pievani@unipd.it

Accepted 6 November 2013; published online 6 December 2013

Abstract
In this paper, we discuss the limits of the traditional view that supernatural beliefs and behaviours are adaptations for social life. We compare it to an alternative hypothesis, according to which supernatural thinking is a secondary effect of cognitive predispositions originally shaped for different adaptive reasons. Finally, we discuss the respective role of such predispositions and socio-cultural factors in shaping and promoting the diffusion of supernatural beliefs.

Keywords
supernatural beliefs, exaptation, animacy detection, agency detection, intuitive dualism, teleology, cognitive predispositions.

1. Introduction

What are the relations between supernatural beliefs, in particular religious beliefs, and morality? According to a traditional view, beliefs in and commitment to supernatural entities emerged because they favoured cooperation and group cohesion. In this paper, we discuss the limits of this view, and we compare it to an alternative one. We start by referring to evidence that religious belief systems do not necessarily favour prosocial behaviours, and that implicit moral evaluations emerge in infancy before the acquisition of religious beliefs. Next, we defend an alternative view according to which supernatural belief systems are secondary effects of cognitive predispositions originally shaped for different adaptive reasons. In particular, we analyse tendencies that emerge early in childhood and might have been advantageous in the process of adaptation to the social and physical world. Finally, we discuss the respective role of such tendencies and socio-cultural factors in shaping supernatural beliefs and promoting their diffusion.

2. Prosocial behaviours and supernatural beliefs: The limits of the traditional view

The thesis that religious beliefs and practices offer advantages to social life is not new (e.g., Durkheim, 1915/1995). Recently, it has enjoyed novel popularity. Basing on evolutionary assumptions, a new version of the thesis argues that religious beliefs are biological adaptations that were directly selected to improve altruism, cooperation and social cohesion within groups competing with other groups (e.g., Wilson, 2003; Bering, 2006). Indeed, creating and maintaining beliefs in supernatural entities who punished cheating and other sorts of non-social behaviour may have favoured social cohesion by increasing the costs of defection, the trust between group members and by sharpening the differences between groups. On the theoretical side, the thesis presents the same problems as other traditional approaches to religious phenomena: It cannot explain why natural selection shaped religious beliefs rather than other cognitive or social means to strengthen social cohesion and cooperation. Moreover, the thesis implies group selection mechanisms. One main problem with group selection is its still weak empirical basis: apart from specific cases already discussed by Darwin (1859), group-selection processes demand very restrictive conditions (e.g., Pievani, 2013a).

If one does not consider such general issues, however, two sets of findings appear to support the view that religious beliefs and practices favour social life. First, in some areas of the world, believers tend to be better citizens than non-believers. In particular, in the US, those who adhere to some religious faith donate more (time and money) to good causes, like charities, than non-believers (Brooks, 2006). From survey data of this sort, some authors have inferred that the religious belief systems may encourage prosocial behaviour more than secular ones (e.g., Haidt, 2007). There are some problems with this conclusion. On the one hand, it does not consider the difficulties that non-believers may experience in contexts in which they are the most despised social minority (Edgell et al., 2006) and at the same time they are asked to contribute to good causes by religious associations. On the other hand, the conclusion neglects correlations that go in the opposite direction. Indeed, various indices of social pathologies, like homicide and youth suicide, are much higher in the US than in more secularized democracies like France or Sweden (Paul, 2005). The same tendencies emerge when one compare areas of the same country presenting different levels of secularization. For example, the more secularized regions of the Centre-North of Italy present both higher economic standards and higher stock of social capital (measured by indicators like associational life, newspaper readership and electoral turnout) than the less secularized regions of the South (Putnam, 1993). In sum, at the population level at least, religious faith is not always positively correlated with pro-social tendencies.

Second, evidence exists that the activation of religious thoughts may encourage prosocial behaviour. Specifically, reminders of moralizing deities, who watch and judge humans, increase generosity (e.g., Sharif & Norenzayan, 2007) and decrease cheating (Mazar et al., 2008). Religious reminders, however, are not necessary to increase social behaviour: reminders of secular authorities and norms produce similar effects (Sharif & Norenzayan, 2007; Mazar et al., 2008). More importantly, secular reminders may countervail religious ones. For example, seminary students who had read the Good Samaritan parable (i.e., the exemplification of compassion toward human suffering) were less likely to help a victim, if the experimenter told them to hurry to another place (Darley & Batson, 1973). In other words, believers who were reminded of religious moral values were less likely to behave pro-socially, if they received a conflicting command by a secular authority. Finally, as illustrated by Norenzayan (2014, this issue), moralizing deities are a recent cultural product that emerged when societal size and complexity increased: Watchful gods were not necessary in small-scale societies, where kin selection, reciprocal altruism and social norms supported cooperation. Notice that even in relatively large societies gods may be more concerned with their own welfare than with human deeds. For example, according to Pliny the Younger, when the Vesuvius erupted, the inhabitants of Pompeii believed that their Gods had left the ending world (Veyne, 2005). In sum, believing in and committing to moralizing gods may enhance pro-social behaviour. Religious morality, however, appears to be a secondary and recent cultural development.

In the past, many authors have claimed that religious beliefs are necessary to shape moral intuitions and to improve human societies. Notably, leaders of democratic societies, ranging from Benjamin Franklin, (Isaacson, 2003) to the former French president Nicolas Sarkozy (2004), have endorsed this claim, possibly the main source of believers' distrust of atheists (Norenzayan, 2014, this issue). Recent research lines, however, have contradicted this idea. Indeed, preverbal or barely verbal children react empathically toward the distress and pain of other individuals (Zahn-Waxler et al., 1992). Likewise, infants make implicit moral evaluations, on the basis of which they seem to prefer agents that help and do not damage others agents (Hamlin et al., 2007). And, as discussed in many papers of the present issue, non-human animals too exhibit similar behaviours (for a review, see De Waal, 2013). In sum, contrary to the traditional view, individuals manifest pro-social tendencies before receiving religious or secular education about moral issues. Along with the evidence reviewed by Norenzayan (2014, this issue), these findings lead to the conclusion that religious beliefs were not selected to enhance prosocial behaviour.

3. Supernatural beliefs as by-products

Unlike the traditional view, an alternative view assumes that supernatural beliefs are not adaptations shaped by natural selection to accomplish specific goals, but by-products of some direct adaptations. Different versions of this view have been proposed (e.g., Guthrie, 1993; Sperber, 1996; Boyer, 2001; Atran, 2002; Barrett, 2004; Bloom, 2004). They all rest on the hypothesis that apparently ubiquitous supernatural beliefs, like beliefs in immaterial souls, magical entities and miracles, emerged out of some cognitive predispositions that have evolved for understanding the physical and social world. From a theoretical perspective, the hypothesis assumes that structures and behaviours developed in previous selection contexts are reused for new functions under different circumstances and in novel ecological niches. This process, currently known as exaptation (Gould & Vrba, 1982), was inferred by Darwin himself in the sixth edition of *The Origin of Species* and then named 'pre-adaptation' (Pievani, 2003). Evidence of various sorts (e.g., morphological, molecular, paleo-bio-geographic) may help distinguish the adaptive versus exaptive nature of the evolution of various traits and behaviours (Pievani & Serrelli, 2011). For example, converging evidence supports the hypothesis that prosocial behaviours were shaped by exaptation processes (Pievani, 2011). Exaptation processes do not imply a suspension of functionality and should be distinguished by the more challenging notion of 'spandrels', whereas a trait originally not shaped by natural selection is then functionally coopted. 'By-product' means here a functional cooptation as side effect of previously evolved adaptive traits.

As for supernatural beliefs formation, Darwin (1871) posited a link between the advantageous tendency to detect agency in natural phenomena and the later emergence of animistic beliefs. The results of recent studies from a variety of fields (e.g., developmental and cognitive psychology, cognitive anthropology and neurosciences) offer an empirical support to Darwin's idea and, more generally, to the by-product view of supernatural beliefs. In what follows, we provide an overview of these lines of evidence.

3.1. Detecting causes, agents and designs

Beginning early in infancy, humans have a strong inclination to trace causal patterns and relations between events. Recently, Mascalzoni et al. (2013) found that newborns between 7 and 72 h after their birth preferentially look at a disc moving on a computer screen and hitting a second disc in such a way as to give the impression of setting it in motion, than at the same discs moving with similar trajectories, but not conveying the impression that the collision with the first disc caused the second to move. In other words, newborns are tuned to causality events as opposed to events with similar kinematic characteristics but not perceived as causally connected. The predisposition to grasp cause-effect relations on the basis of the spatio-temporal correlation between events, however, is not sufficient in itself to engender supernatural beliefs.

A natural correlate of the causal stance is the tendency to infer the presence of possible, sometimes invisible, causes. The behaviours of agents are a particular category of causes. Unlike inert objects, which are moved by external forces, agents can move of their own and make other entities move. Moreover, unlike inert objects, animated ones *act* rather than simply move in arbitrary ways, namely, their actions are geared towards goals. Infants and young children appear to have a strong inclination to reason about entities as agents, in the presence even of the slightest traces of animacy. Thus, they can treat geometrical shapes as agents. For example, 3-month-olds prefer to watch a video in which two discs (one red and one blue) move as though one were chasing the other to a similar video in which the same discs move about in a casual manner (Rochat et al., 1997). Moreover, infants appear to draw sophisticated inferences about the motion of geometrical, but apparently animated, shapes. For example, Rochat et al. (2004) showed nine-month-olds a video in which a red disc chased a blue disc. When infants were habituated (i.e., their looking time decreased), the experimenters presented them a similar video in which the blue disc chased the red one. If infants had habituated just to the chasing motion, they should remained habituated to the new video, since it presented the same perceptual features than the previous one. By contrast, if infants had assigned different roles to the two discs in motion, they should regained interest in the scene, because the discs had changed their respective role. Indeed, infants looked longer at the new video. In sum, from the age of nine months onwards, infants appear to notice role changes in scenes in which geometrical shapes enact chasing motion. Finally, infants infer the presence of agents even when they are not visible. For example, Saxe et al. (2005) showed 10- and 12-month-olds a scene in which an invisible person threw a beanbag over a wall and onto a stage. When infants were habituated, a human arm entered the stage and stopped with the hand in the centre of it, either from the same side from which the beanbag emerged in motion or from the opposite side. Infants looked longer the scene when the harm came from the side opposite the beanbag. In other words, by the end of their first year, infants are able to draw inferences about the invisible causal agent of the motion of an inanimate object.

The reviewed evidence shows that infants have an early and strong tendency to attribute agency based on minimal cues. Indeed, all humans, including adults, seem to be highly sensitive to signs of agency, in particular, of human agency, so much so that they tend to attribute intentions where all that really exists is accident (Boyer, 2001). Likewise, humans are pattern seekers, so much so that they tend regard regularities, like a series of six boys born in sequence at a hospital, not as accident but as the product of mechanical causality or of someone's intention (Kahneman, 2011).

Most of this machinery for animacy and agency detection seems to be operative in the brain of other animals as well (for reviews, see Vallortigara, 2012a, b). In fact, Darwin, who was probably the first to connect agency detection and the emergence of supernatural beliefs, illustrated his idea by means of an episode concerning a dog:

"The tendency in savages to imagine that natural objects and agencies are animated by spiritual or living essences, is perhaps illustrated by a little fact which I once noticed: my dog, a full-grown and very sensible animal, was lying on the lawn during a hot and still day; but at a little distance a slight breeze occasionally moved an open parasol, which would have been wholly disregarded by the dog, had any one stood near it. As it was, every time that the parasol slightly moved, the dog growled fiercely and barked. He must, I think, have reasoned to himself in a rapid and unconscious manner, that movement without any apparent cause indicated the presence of some strange living agent, and that no stranger had a right to be on his territory. The belief in spiritual agencies would easily pass into the belief in the existence of one or more gods. For savages would naturally attribute to spirits the same passions, the same love of vengeance or simplest form of justice, and the same affections which they themselves feel" (Darwin, 1871, p. 67).

The above passage indicates that Darwin regarded the faculty of being sensitive to agency cues as an advantageous one. Recently, many authors have attributed the tendency to detect agency, even when it is inappropriate to do so, to a rational strategy: better safe than sorry (e.g., Guthrie, 1993; Boyer, 2001; Barrett, 2004). Indeed, the cost of over-detecting agency (e.g., an animal who reacts to a parasol's move, and it is just the wind, does not lose anything) is lower than the cost of under-detecting it (e.g., an animal who does not react, and there is an enemy there, might lose a lot). As indicated above, the agency detector is hyperactive in our species, possibly because of the high degree of sophistication of our social life (Humphrey, 1984).

What could be the consequences of such a tendency? Following Darwin, modern scholars have argued that it is the source of animistic religious beliefs (e.g., Guthrie, 1993). It is likely, however, that it also play a major role in the emergence and diffusion of two other fundamental features of supernatural thinking: design stance and teleological stance. The first feature refers to the human propensity to use the designer's intention to categorize objects and to infer their intended functions, ignoring the actual, and possibly messy, details of their physical constitution (Dennett, 1987). Beginning early in childhood, humans exhibit such a propensity. For example, 3-year-olds are more likely to call an object an hat, when they are informed that a piece of newspaper has been shaped with this intention by somebody, than when they are informed that a piece of newspaper has been incidentally shaped in this way (Gelman & Bloom, 2000). Unlike non-human primates, humans aptly use the design stance in learning how to interact with objects. When children are given a new tool, they use the intention of the individual who presents it, in order to understand how it works. By contrast, chimpanzees tend to rely on its perceptual features (Horner & Whiten, 2005). Besides emerging early in childhood, the design stance persists in adulthood across cultures, informing the acquisition and representation of knowledge about artefacts. Even individuals who live in nonindustrial, technologically sparse cultures are sensitive to information about design function. German & Barrett (2005) asked a group of adolescents of the Shuar, an hunter-horticulturalist people of Ecuadorian Amazonia, to tackle problems whose solution implies the atypical use of an artefact (e.g., using a spoon to bridge a gap between two objects). When the design function of the target artefact was not primed during problem presentation (e.g., a spoon was presented alongside other objects), participants performed better than when the design function was primed (e.g., the spoon was presented inside a cup full of rice). In sum, information about the design function constrains problem solving just as it does with individuals living in technologically rich cultures (Duncker, 1945). Taken together, these findings illustrate the adaptive value of the intuitive design stance and its role in human evolution. Without such an advantageous heuristic, our ancestors would probably not have been able to surpass the technologically poor cultures of the other primates, in which the customary or habitual use of tools ranges from five to at most twenty (Whiten et al., 2001). In sum, it is plausible to associate the design stance to the early emergence of technological abilities in the history of humankind (Ambrose, 2001).

Just as humans tend to over-attribute agency, they also tend to over-attribute design. Young children, for example, attribute purpose not only to artefacts, but also to living ("The tiger is made for being seen at a zoo") and non-living natural objects ("This prehistorical rock was pointy so that animals could scratch on it"). Children's tendency to use teleological explanations of this sort, rather than physical-causal explanations, decreases with formal education (Kelemen, 2003), but it may reappear in adult inferences. In particular, Alzheimer patients with degraded semantic memories (Lombrozo et al., 2007) and adults with minimal schooling (Casler & Kelemen, 2008) display teleological intuitions about natural entities and their properties. A recent study, however, has provided striking evidence that the teleological stance is a lifelong cognitive default. Kelemen et al. (2013) asked a group of professional physical scientists at high-ranking American research universities to judge sentences as true or false. Some sentences were false teleological explanations of natural phenomena (e.g., The sun makes light so that plants can photosynthesise). Others were incongruous teleological explanations concerning artefacts (e.g., Window blinds have slats so that they can capture dust). In a control condition, in which participants had no time limits to answers, they did not endorse false explanations of any sorts. However, when participants had to judge explanations at speed — a condition that hinders inhibition of automatic explanatory reactions — they showed higher acceptance of false teleological explanations. In sum, the strongest possible test of the teleological stance hypothesis yielded clear-cut results: Even individuals who have relevant scientific knowledge and explicitly reject teleological explanations of nature in their professional life, endorse them when their cognitive resources are taxed. This set of findings indicates that teleological intuitions emerge early in childhood and may be inhibited by later scientific education, but are never fully eliminated.

If there is an intuitive tendency to discern design in nature, is there also a tendency to infer the presence of a Designer? In other words, is there an intuitive equivalent of the design argument? In Paley's (1802/2006) famous version, the argument is that if one observes the complexity of a watch, one is lead to infer that it was purposefully designed by a maker. Likewise, if one observes the complexity of the nature, one is lead to infer that it was purposefully designed by a Supreme Maker. Formally, the design argument implies the ability to relate observations and potential explanations via the vehicle of likelihoods (Sober, 2004) or probabilities (DeCruz & De Smedt, 2010). Preverbal infants seem to have an implicit understanding of the principles that connect observations and hypotheses (Teglas et al., 2011). And pre-schoolers manifest such an understanding in an explicit, albeit qualitative way, in their judgments and choices (Girotto & Gonzalez, 2008; Gonzalez & Girotto, 2011). Finally, children not only have the logical prerequisites for the design argument, but actually seem to endorse it. When nine-year-olds are explicitly asked to select one explanation about the origin of animals, they prefer creationism over natural selection, regardless of whether they have been raised by a creationist or a secular community (Evans, 2001).

Some authors have attributed the enduring appeal of the design argument to cultural factors, like the pervasive influence of the Aristotelian teleology (Wattles, 2006). The evidence reviewed in this section suggests a different conclusion: Unlike the a-teleological and purposeless evolutionary theory, the design argument rests on evolved tendencies of the human mind. Thus, it maintains its intuitive attractiveness one and half century after the publication of Darwin's theory (Pievani, 2013b).

3.2. Intuitive dualism

Data from developmental psychology, evolutionary psychology and neurosciences suggest that, beginning in early infancy, humans have two distinct cognitive systems: naive psychology and naive physics (Bloom, 2004). Infants deal with information concerning animate entities in a different way from physical entities. For example, they usually lose interest in a moving object that stops its trajectory, but are afraid of a person who suddenly stops interacting with them and keeps still. Humans and non-human primates seem to have similar naïve physics. Humans, however, possess a much more sophisticated naïve psychology (Povinelli & Vonk, 2003), possibly, once again, because of the complexity of their social life.

Recently, many authors have argued that the incommensurable outputs of these two cognitive systems have generated, as an evolutionary by-product, an intuitive Cartesianism, namely, a basic ontological distinction between spirit and matter (e.g., Bloom, 2004). Infants and young children manifest dualistic tendencies. Consider the above described study on secret agent detection (Saxe et al., 2005). Infants looked longer when the human harm came from the opposite side of the scene than when it came from the same side as the beanbag. However, infants' looking times were the same for a toy train that came from the same side as the beanbag and for a toy train that came from the opposite one. In other words, infants drew different causal inferences about animate versus inanimate entities: A human harm is a plausible causal agent of an object's motion but not a train-toy. Moreover, infants attribute the power to produce order from chaos only to animate entities. Newman et al. (2010) showed 12-month-olds a scene in which a ball moved towards a screen concealing a pile of casually arranged building blocks. Infants looked longer when the subsequent lowering of the screen revealed that the collision with the ball had arranged the blocks in order than when the ball scattered a group of previously ordered blocks. However, if the ball had a pair of eyes, which turned it into an animated agent, infants' looking time did not vary with the outcome of the collision, as if they assumed that agents, but not inanimate objects, could produce order. Notice that a similar intuition inspired the first versions of the design argument. In the *Nature of Gods*, Cicero rejected the atomistic theory that the fortuitous collision of atoms generated the nature, arguing that chance could not produce order. One who believes the atomistic theory, Cicero argued, may as well believe that the throwing of a great quantity of the 21 letters of the Latin alphabet could fall into such order as to form an intelligible text.

Older children express their dualistic views more explicitly. For example, pre-schoolers attribute an individual's identity to her mind rather than to her brain. If a girl is magically transformed into a seal but her mind remains unchanged, she retains her knowledge and memory. However, if the transformation concerns both her body and her brain, she gets the knowledge and memory of the seal (Corriveau et al., 2005).

The crucial consequence of the intuitive dualism is that individuals are inclined to conceive bodies without spirits and, more importantly, spirits without bodies, like divinities, ghosts, demons, immaterial souls, and so on. Moreover, the intuitive dualism allows individuals to believe in the existence of an afterlife. For example, pre-schoolers believe that, unlike its body, the mental properties of a dead mouse continue to work, so that it can hold desires and think thoughts. Importantly, this tendency is stronger among younger children, suggesting that it is a natural intuition that can be over-ridden only in later childhood, if at all (Bering, 2006).

4. From cognitive predispositions to cultural scaffoldings

The outlined view of religiosity by accident yields predictions that are em-pirically testable. In what follows, we consider three of them. First, if be-lieving in personified supernatural entities is a by-product of adaptive socio-cognitive propensities, like representing others' mental states, then individ-uals who possess these propensities in a weaker form should also possess a weaker tendency to believe in deities having feelings, desires and beliefs. Norenzayan et al. (2012) have corroborated this prediction, showing that the autistic syndrome, which entails difficulties in representing and inferring the mental states of others, is associated with a weakened tendency to believe in personified deities. They also found a correlation between males' weaker ca-pacity to adequately represent others' mental states and their weaker propen-sity towards religiosity. Second, if religious beliefs are a by-product of social cognition, then thinking about or praying to personified deities should acti-vate the same brain areas as thinking about other individuals. Brain-imaging studies have corroborated both predictions (e.g., Harris et al., 2009; Ka-pogiannis et al., 2009; Schjødt et al., 2009). Third, if supernatural beliefs are supported by intuitive cognitive processes, then more analytical ways of thinking ought to weaken them. Shenhav et al. (2012) asked a group of participants to describe a positive result they had obtained through carefully reasoning, and then to fill in a questionnaire on religiosity. These participants reported weaker belief in God compared to participants asked to describe a positive result they had attained through intuition. Gervais & Norenzayan (2012) have reached similar results by activating analytical thinking in an even more subtle way. Some participants were asked to rearrange a series of words, one of which related to analytical thinking (e.g., reason, ponder, ratio-nal), and then to answer a few questions about religiosity. These individuals showed less adherence to religious beliefs than those asked to rearrange neu-tral words (e.g., hammer, jump, shoes). Strategies for analytical thinking, therefore, can weaken the intuitions underlying religious beliefs.

Granted that the reviewed intuitions are necessary preconditions to en-gender the emergence of supernatural thinking, are they also sufficient to produce it? Some authors would answer 'yes', claiming that supernatural beliefs arise effortlessly from the cognitive tendencies of the human mind with little or no cultural scaffolding (e.g., Barrett, 2004; Bering, 2006). This claim seems to suffer from the same difficulty as the traditional views in that it cannot explain why those specific beliefs have been selected. As Gervais et al. (2011) pointed out, in principle, there are countless supernatural entities (e.g., divinities of other faiths, folk tales characters) that have features (e.g., being invisible) and powers (e.g., flight) that make them attention arresting, memorable and easily communicable. In other words, there are countless 'relevant mysteries', as Sperber (1996) called them, that could become ob-ject of worship. Yet, individuals come to believe in and to sincerely commit to only a fraction of these mysteries. In particular, individuals tend to pas-sionately believe only in the specific supernatural entities that are prevalent in the cultural context in which they live. Thus, if one wants to explain the persistence and diffusion of a specific set of supernatural beliefs, one should consider the role of the social mechanisms of persuasion, like social proof, authority and liking (Cialdini, 2001). Individuals are more likely to come to believe in and adhere to supernatural beliefs that are supported by most other members in their community, by prestigious members or by members who display costly signs of their own adherence (Gervais et al., 2011).

There is an alternative interpretation of the claim that supernatural beliefs emerge naturally from cognitive predispositions. One might argue that only few, basic beliefs require little or no cultural input. Under this interpretation, the belief in the existence of an afterlife is a natural consequence of human intuitive dualism, and the belief in a supernatural maker of the living species is a natural consequence of the design and teleological stances. The available evidence, however, does not seem to corroborate such an interpretation. As for the belief in an afterlife, the above mentioned study on the dead mouse's traits appears to support the claim that this belief is a default cognitive stance that does not need cultural learning (Bering, 2006). Indeed, younger children appeared to attribute persistent mental properties to the dead mouse more than older ones. Other studies, however, have reported contrasting results. Harris & Giménez (2005) asked Spanish 7- and 11-year-olds questions about a grandparent's death, which was described either in a secular or in a religious frame. Notice that, unlike the dead mouse study, in this case children were asked to reason about the death of a real individual, not about a fictional character. The results showed that older, rather than younger, children, were more likely to claim than mental processes continue after death, especially when the death was framed in religious terms. This result suggests that afterlife beliefs do not appear early in childhood, and their emergence may be favoured by religious enculturation. Astuti & Harris (2008) replicated and extended this finding. They investigated death-related beliefs in adults and children of the Vezo, a group of people who live in rural Madagascar and frequently perform rituals to ingratiate dead relatives. Youngest children (5-year-olds) did not exhibit a reliable pattern in their conception of death. Older children (7-year-olds) claimed that both physical and mental processes ceased at death. Still older children (12-year-olds) and adults claimed that mental processes continued after death. In sum, for both Western and non-Western children, afterlife beliefs seem to emerge late in childhood. This finding cannot be easily explained by the hypothesis that believing in the existence of an afterlife is a default stance. By contrast, this finding supports the view that religious enculturation and, more generally, socio-cultural factors favour the emergence of supernatural beliefs.

As for the belief in a Supreme Designer, the existing evidence does not support the view that there are basic creationist intuitions. Consider the above mentioned study on animal-origin beliefs in children (Evans, 2001). It found that US Midwestern 9-year-olds endorsed creationism, regardless of the community (fundamentalist vs. non-fundamentalist) in which they had been raised. However, it also found that younger children (about 7-year-olds) did not have clear intuitions about the origins of animals: They were equally likely to accept creationist, evolutionist or spontaneous generation accounts. Moreover, it also found that older children (11-year-olds) endorsed the beliefs of their parents. In particular, most children who did not attend fundamentalist churches and fundamentalist schools believed in evolution. Finally, consider again the study on the teleological intuitions of Alzheimer patients (Lombrozo et al., 2007). It showed that these patients tend to use the same teleological explanations as young children. However, it also showed that they were not more likely to invoke a supernatural designer than healthy participants. In sum, the current evidence suggests that there is a strong and enduring teleological stance, which emerges early in infancy and never completely disappears (Kelemen et al., 2012), but it does not suggest that there is also an independent creationist stance.

5. Conclusions

In this paper, we have reviewed some of the studies that support the by-product account of supernatural thinking. The discovery of basic predispositions (i.e., causal, design and teleological stance, intuitive dualism) has improved our understanding of the cognitive and evolutionary origins of this form of thinking. Basic predispositions, however, may explain how individuals represent supernatural entities, but they cannot explain how individuals come to believe in and to commit to some of these entities. To address these larger questions, one has to take into account the role of cultural factors and persuasion mechanisms in shaping and promoting the diffusion of some specific beliefs. An integrated perspective of this kind is likely to produce a proper natural investigation of supernatural thinking.

Acknowledgements

V.G. was supported by Swiss & Global, Ca' Foscari Foundation and the Italian Ministry of Research (grant PRIN2010-RP5RNM). G.V. was supported by an ERC Advanced Grant (PREMESOR ERC-2011-ADG_20110406).

References

Ambrose, S.H. (2001). Paleolithic technology and human evolution. — Science 291: 1748-1753.

Astuti, R. & Harris, P.L. (2008). Understanding mortality and the life of the ancestors in rural Madagascar. — Cogn. Sci. 32: 713-740.

Atran, S. (2002). In Gods we trust. The evolutionary landscape of religion. — Oxford University Press, Oxford.

Barrett, J.L. (2004). Why would anyone believe in God? — Altamira Press, Lanham, MD.

Bering, J.M. (2006). The folk psychology of souls. — Behav. Brain Sci. 29: 453-462.

Bloom, P. (2004). Descartes' baby: how the science of child development explains what makes us human. — Basic Books, New York, NY.

Boyer, P. (2001). Religion explained. — Basic Books, New York, NY.

Brooks, A.C. (2006). Who really cares: the surprising truth about compassionate conservatism. — Basic Books, New York, NY.

Casler, K. & Kelemen, D. (2008). Developmental continuity in teleofunctional explanation: reasoning about nature among Romanian Romani adults. — J. Cogn. Dev. 9: 340-362.

Cialdini, R. (2001). Influence: the psychology of persuasion. — Alley & Bacon, Boston, MA.

Corriveau, K.H., Pasquini, E.S. & Harris, P. (2005). "If it's in your mind, it's in your knowledge": children's developing anatomy of identity. — Cogn. Dev. 20: 321-340.

Darley, J. & Batson, C.D. (1973). From Jerusalem to Jericho: a study of situational and dispositional variables in helping behaviour. — J. Personal. Soc. Psychol. 27: 100-108.

Darwin, C.R. (1859). The origins of species. — John Murray, London.

Darwin, C.R. (1871). The descent of man, and selection in relation to sex. — John Murray, London.

De Cruz, H. & De Smedt, J. (2010). The innateness hypothesis and mathematical concepts. — Int. Rev. Philos. 29: 3-13.

de Waal, F.B.M. (2013). The bonobo and the atheist. In search of humanism among the primates. — W.W. Norton, New York, NY.

Dennett, D. (1987). The intentional stance. — MIT Press, Cambridge, MA.

Duncker, K. (1945). On problem-solving. — Psychol. Monogr. 58(5, Whole No. 270).

Durkheim, E. (1995/1915). The elementary forms of the religious life. — The Free Press, New York, NY.

Edgell, P., Gerteis, J. & Hartmann, D. (2006). Atheists as "Other": moral boundaries and cultural membership in American society. — Am. Sociol. Rev. 71: 211-234.

Evans, E.M. (2001). Cognitive and contextual factors in the emergence of diverse belief systems: creation versus evolution. — Cogn. Psychol. 42: 217-266.

Gelman, S.A. & Bloom, P. (2000). Young children are sensitive to how an object was created when deciding what to name it. — Cognition 76: 91-103.

German, T.P. & Barrett, H.C. (2005). Functional fixedness in a technologically sparse culture. — Psychol. Sci. 16: 1-5.

Gervais, W.M. & Norenzayan, A. (2012). Analytic thinking promotes religious disbelief. — Science 336: 493-496.

Gervais, W.M., Willard, A.K., Norenzayan, A. & Henrich, J. (2011). The cultural transmission of faith: why innate intuitions are necessary, but insufficient, to explain religious belief. — Religion 41: 389-410.

Girotto, V. & Gonzalez, M. (2008). Children's understanding of posterior probability. — Cognition 106: 325-344.

Gonzalez, M. & Girotto, V. (2011). Combinatorics and probability: six-to-ten-year-olds reliably predict whether a relation will occur. — Cognition 120: 372-379.

Gould, S.J. & Vrba, E.S. (1982). Exaptation, a missing term in the science of form. — Paleobiology 8: 4-15.

Guthrie, S. (1993). Faces in the clouds: a new theory of religion. — Oxford University Press, Oxford.

Haidt, J. (2007). Moral psychology and the misunderstanding of religion. Edge. The third culture. — Available online at: http://www.edge.org/3rd_culture/haidt07/haidt07_index.html

Hamlin, J.K., Wynn, K. & Bloom, P. (2007). Social evaluation by preverbal infants. — Nature 450: 557-559.

Harris, P.L. & Gimenez, M. (2005). Children's acceptance of conflicting testimony: the case of death. — J. Cogn. Cult. 5: 143-164.

Harris, S., Kaplan, J.T., Curiel, A., Bookheimer, S.Y. & Iacoboni, M. (2009). The neural correlates of religious and nonreligious belief. — PLoS ONE 4(10): e0007272, DOI:10.1371/journal.pone.0007272.

Horner, V. & Whiten, A. (2005). Causal knowledge and imitation/emulation switching in chimpanzees (Pan troglodytes) and children (Homo sapiens). — Anim. Cogn. 8: 164-181.

Humphrey, N. (1984). Consciousness regained. — Oxford University Press, Oxford.

Isaacson, W. (2003). Benjamin Franklin: an American life. — Simon & Shuster, New York, NY.

Kahneman, D. (2011). Thinking, fast and slow. — Allen Lane, London.

Kapogiannis, D., Barbey, A.K., Su, M., Zamboni, G., Krueger, F. & Grafman, J. (2009). Cognitive and neural foundations of religious belief. — Proc. Natl. Acad. Sci. USA 106: 4876-4881.

Kelemen, D. (2003). British and American children's preferences for teleo-functional explanations of the natural world. — Cognition 88: 201-221.

Kelemen, D., Rottman, J. & Seston, R. (2012). Professional physical scientists display tenacious teleological tendencies: purposed-based reasoning as a cognitive default. — J. Exp. Psychol., in press, DOI:10.1037/a0030399.

Lombrozo, T., Kelemen, K. & Zaitchik, D. (2007). Inferring design: evidence of a preference for teleological explanations in patients with Alzheimer's disease. — Psychol. Sci. 18: 999-1006.

Mascalzoni, E., Regolin, L., Vallortigara, G. & Simion, F. (2013). The cradle of causal reasoning: newborns' preference for physical causality. — Dev. Sci. 16: 327-335.

Mazar, N., Amir, O. & Ariely, D. (2008). The dishonesty of honest people: a theory of self-concept maintenance. — J. Mark. Res. 45: 633-644.

Newman, G., Keil, F.C., Kuhlmeier, V.A. & Wynn, K. (2010). Early understandings of the link between agents and order. — Proc. Natl. Acad. Sci. USA 107: 17140-17145.

Norenzayan, A. (2014). Does religion make people moral? — Behaviour 151: 365-384.

Norenzayan, A., Gervais, W.M. & Trzesniewski, K. (2012). Mentalizinig deficits constraint belief in a personal God. — PLoS ONE 7: e36880.

Paley, W. (2006/1802). Natural theology. — Oxford University Press, Oxford.

Paul, G. (2005). Cross-national correlations of quantifiable societal health with popular religiosity and secularism in the prosperous democracies: a first look. — J. Rel. Soc. 7: 20-26.

Pievani, T. (2003). Rhapsodic evolution: essay on exaptation and evolutionary pluralism. World futures. — J. Gen. Evol. 59: 63-81.

Pievani, T. (2011). Born to cooperate? Altruism as exaptation, and the evolution of human sociality. — In: Origins of cooperation and altruism (Sussman, R.W. & Cloninger, R.C., eds). Springer, New York, NY, p. 41-61.

Pievani, T. (2013a). Individuals and groups in evolution: Darwinian pluralism and the multi-level selection debate. — J. Biosci. 38: DOI:10.1007/s12038-013-9345-4.

Pievani, T. (2013b). Intelligent design and the appeal of teleology: structure and diagnosis of a pseudoscientific doctrine. — Paradigmi: DOI:10.3280/PARA2013-001010.

Pievani, T. & Serrelli, E. (2011). Exaptation in human evolution: how to test adaptive vs exaptive evolutionary hypotheses. — J. Anthropol. Sci. 89: 1-15.

Povinelli, D.J. & Vonk, J. (2003). Chimpanzee minds. Suspiciously humans? — Trends Cogn. Sci. 7: 157-160.

Putnam, R.D. (1993). Making democracy work: civic traditions in modern Italy. — Princeton University Press, Princeton, NJ.

Rochat, P., Morgan, R. & Carpenter, M. (1997). Young infants' sensitivity to movement information specifying social causality. — Cogn. Dev. 12: 536-561.

Rochat, P., Striano, T. & Morgan, R. (2004). Who is doing what to whom? Young infants' developing sense of social causality in animated displays. — Perception 33: 355-369.

Sarkozy, N. (2004). La république, les religions, l'espérance. — Les Editions du Cerf, Paris.

Saxe, R., Tenenbaum, J.B. & Carey, S. (2005). Secret agents: inferences about hidden causes by 10- and 12-month-old infants. — Psychol. Sci. 16: 995-1001.

Schjødt, U., Stodkilde-Jorgensen, H., Geertz, A.W. & Roepstorff, A. (2009). Highly religious participants recruit areas of social cognition in personal prayer. — SCAN 4: 199-207.

Shenhav, A., Rand, D. & Greene, D. (2012). Divine intuition: cognitive style influences belief in God. — J. Exp. Psychol.: Gen. 141: 423-428.

Sober, E. (2005). The design argument. — In: The Blackwell guide the philosophy of religion (Mann, W.E., ed.). Blackwell, Oxford, p. 117-147.

Sperber, D. (1996). Explaining culture: a naturalistic approach. — Blackwell, Oxford.

Téglás, E., Vul, E., Girotto, V., González, M., Tenenbaum, J.B. & Bonatti, L.L. (2011). Pure reasoning in 12-month-old infants as probabilistic inference. — Science 332: 1054-1058.

Vallortigara, G. (2012a). Aristotle and the chicken: animacy and the origins of beliefs. — In: The theory of evolution and its impact (Fasolo, A., ed.). Springer, New York, NY, p. 189-200.

Vallortigara, G. (2012b). Core knowledge of object, number, and geometry: a comparative and neural approach. — Cogn. Neuropsychol. 29: 213-236.

Wattles, J. (2006). Teleology past and present. — Zygon: J. Relig. Sci. 41: 445-464.

V. Girotto et al.

Whiten, A., Goodall, J., McGrew, W.C., Nishida, T., Reynolds, V., Sugiyama, Y., Tutin, C.E.G., Wrangham, R.W. & Boesch, C. (2001). Charting cultural variations in chimpanzees. — Behaviour 138: 1481-1516.

Wilson, D.S. (2003). Darwin's cathedral: evolution, religion, and the nature of society. — University of Chicago Press, Chicago, IL.

Zahn-Waxler, C., Radke-Yarrow, M., Wagner, E. & Chapman, M. (1992). Development of concern for others. — Dev. Psychol. 28: 126-136.

Index

Printed in the United States
by Baker & Taylor Publisher Services